Communications in Computer and Information Science   **577**

Commenced Publication in 2007
Founding and Former Series Editors:
Alfredo Cuzzocrea, Dominik Ślęzak, and Xiaokang Yang

More information about this series at http://www.springer.com/series/7899

Dominique de Werra · Greg H. Parlier
Begoña Vitoriano (Eds.)

Operations Research and Enterprise Systems

4th International Conference, ICORES 2015
Lisbon, Portugal, January 10–12, 2015
Revised Selected Papers

 Springer

Editors
Dominique de Werra
EPFL SB-DO
Lausanne
Switzerland

Begoña Vitoriano
Complutense University
Madrid
Spain

Greg H. Parlier
MAS of INFORMS
Madison, AL
USA

ISSN 1865-0929   ISSN 1865-0937   (electronic)
Communications in Computer and Information Science
ISBN 978-3-319-27679-3   ISBN 978-3-319-27680-9   (eBook)
DOI 10.1007/978-3-319-27680-9

Library of Congress Control Number: 2015956372

Printed on acid-free paper

This Springer imprint is published by SpringerNature
The registered company is Springer International Publishing AG Switzerland

Preface

This book includes extended and revised versions of selected papers presented during the 5th International Conference on Operations Research and Enterprise Systems (ICORES 2015), held in Lisbon, Portugal, during January 10–12, 2015. ICORES 2015 was sponsored by the Institute for Systems and Technologies of Information, Control and Communication (INSTICC) and co-sponsored by the Portuguese Association of Operational Research (Apdio).

The purpose of the International Conference on Operations Research and Enterprise Systems is to bring together researchers, engineers, and practitioners interested in both research and practical applications in the field of operations research. Two simultaneous tracks were held, one focused on methodologies and technologies and the other on practical applications in specific areas.

ICORES 2015 received 89 paper submissions from 38 countries across six continents. Of these, 21 % were presented at the conference as full papers. These authors were then invited to submit extended versions of their papers. Each submission was evaluated during a double-blind review by the conference Program Committee. The best 18 papers were selected for publication in this book.

ICORES 2015 also included four plenary keynote lectures from internationally distinguished researchers: Francisco Ruiz, (University of Málaga, Spain), Marc Demange (School of Mathematical and Geospatial Sciences, RMIT University, Australia), Marino Widmer (University of Fribourg, Switzerland), and Bernard Ries (Université Paris-Dauphine, France). We gratefully acknowledge their invaluable contribution as renowned experts in their respective areas. They presented cutting-edge work, thus enriching the scientific content of the conference.

We especially thank all authors whose research and development efforts are recorded here. The knowledge and diligence of our reviewers were also essential to ensure that high-quality papers were presented at the conference and published herein. Finally, our special thanks to all members of the INSTICC team for their indispensable administrative skills and professionalism, both of which contributed to a well-organized, productive, and memorable conference.

September 2015

Begoña Vitoriano
Greg H. Parlier

Organization

Conference Chair

Dominique de Werra — École Polytechnique Fédérale de Lausanne (EPFL), Switzerland

Program Co-chairs

Begoña Vitoriano — Complutense University, Spain
Greg H. Parlier — MAS of INFORMS, USA

Program Committee

Mohamed Abido	KFUPM, Saudi Arabia
A.A. Ageev	Sobolev Institute of Mathematics (Russian Academy of Sciences), Russian Federation
El-Houssaine Aghezzaf	Ghent University, Belgium
Javier Alcaraz	Universidad Miguel Hernandez de Elche, Spain
Maria Teresa Almeida	ISEG, UTL, Portugal
Lyes Benyoucef	Aix-Marseille University, France
Jean-Charles Billaut	Ecole Polytechnique de l'Université François-Rabelais de Tours, France
Christian Blum	IKERBASQUE and University of the Basque Country, Spain
Ralf Borndörfer	Zuse Institute Berlin, Germany
Endre Boros	Rutgers University, USA
Ahmed Bufardi	Ecole Polytechnique Federale de Lausanne, Switzerland
Sujin Bureerat	KhonKaen University, Thailand
Alfonso Mateos Caballero	Universidad Politécnica de Madrid, Spain
José Manuel Vasconcelos Valério de Carvalho	Universidade do Minho, Portugal
Dirk Cattrysse	Katholieke Universiteit Leuven, Belgium
Bo Chen	University of Warwick, UK
John Chinneck	Carleton University, Canada
James Cochran	University of Alabama, USA
Mikael Collan	Lappeenranta University of Technology, Finland
Xavier Delorme	Ecole Nationale Supérieure des Mines de Saint-Etienne, France
Marc Demange	RMIT University, Australia

Clarisse Dhaenens	French National Institute for Research in Computer Science and Control, France
Tadashi Dohi	Hiroshima University, Japan
Nikolai Dokuchaev	Curtin University, Australia
Christophe Duhamel	Université Blaise Pascal, Clermont-Ferrand, France
Pankaj Dutta	Indian Institute of Technology Bombay, India
Gintautas Dzemyda	Vilnius University, Lithuania
Andrew Eberhard	RMIT University, Australia
Ali Emrouznejad	Aston University, UK
Nesim Erkip	Bilkent University, Turkey
Yahya Fathi	North Carolina State University, USA
Muhammad Marwan Muhammad Fuad	University of Tromsø, Norway
Robert Fullér	Obuda University, Hungary
Heng-Soon Gan	The University of Melbourne, Australia
Michel Gendreau	Ecole Polytechnique de Montréal, Canada
Giorgio Gnecco	IMT, Institute for Advanced Studies, Lucca, Italy
Boris Goldengorin	University of Florida, USA
Juan José Salazar Gonzalez	Universidad de La Laguna, Spain
Christelle Guéret	University of Angers, France
Nalan Gulpinar	The University of Warwick, UK
Gregory Z. Gutin	Royal Holloway University of London, UK
Jin-Kao Hao	University of Angers, France
Han Hoogeveen	Universiteit Utrecht, The Netherlands
Chenyi Hu	The University of Central Arkansas, USA
Johann Hurink	University of Twente, The Netherlands
Josef Jablonsky	University of Economics, Czech Republic
Joanna Józefowska	Poznan University of Technology, Poland
Jesuk Ko	Gwangju University, Republic of Korea
Erhun Kundakcioglu	Ozyegin University, Turkey
Philippe Lacomme	Université Clermont-Ferrand 2, Blaise Pascal, France
Sotiria Lampoudi	Liquid Robotics Inc., USA
Dario Landa-Silva	University of Nottingham, UK
Janny Leung	The Chinese University of Hong Kong, Hong Kong, SAR China
Abdel Lisser	The University of Paris-Sud 11, France
Pierre Lopez	LAAS-CNRS, Université de Toulouse, France
Helena Ramalhinho Lourenço	Universitat Pompeu Fabra, Spain
Viliam Makis	University of Toronto, Canada
Arnaud Malapert	Université Nice Sophia-Antipolis CNRS, France
Patrice Marcotte	Université de Montréal, Canada
Concepción Maroto	Universidad Politécnica de Valencia, Spain
Pedro Coimbra Martins	Polytechnic Institute of Coimbra, Portugal
Nimrod Megiddo	IBM Almaden Research Center, USA

Additional Reviewer

Ad Feelders   Universiteit Utrecht, The Netherlands

Invited Speakers

Francisco Ruiz   University of Málaga, Spain
Marc Demange   School of Mathematical and Geospatial Sciences,
                             RMIT University, Australia
Marino Widmer   University of Fribourg, Switzerland
Bernard Ries   Université Paris-Dauphine, France

Contents

Methodologies and Technologies

Distributionally Robust Optimization for Scheduling Problem in Call Centers with Uncertain Forecasts

Mathilde Excoffier$^{(\boxtimes)}$, Céline Gicquel, Oualid Jouini, and Abdel Lisser

Laboratoire de Recherche en Informatique - LRI, 91405 Orsay Cedex, France
mathilde.excoffier@lri.fr

Abstract. This paper deals with the staffing and scheduling problem in call centers. We consider that the call arrival rates are subject to uncertainty and are following independent unknown continuous probability distributions. We assume that we only know the first and second moments of the distribution and thus propose to model this stochastic optimization problem as a distributionally robust program with joint chance constraints. Moreover, the risk level is dynamically shared throughout the entire scheduling horizon during the optimization process. We propose a deterministic equivalent of the problem and solve linear approximations of the Right-Hand Side of the program to provide upper and lower bounds of the optimal solution. We applied our approach on a real-life instance and give numerical results. Finally, we showed the practical interest of this approach compared to a stochastic approach in which the choice of the distribution is incorrect.

Keywords: Distributionally robust optimization · Stochastic programming · Joint chance constraints · Mixed-integer linear programming · Staffing · Shift-scheduling · Call centers · Queuing systems

1 Introduction

Call centers are the main interface between the firms and their clients: in the U.S. in 2002, call centers represent 70 % of all business interactions [3]. Whether it be for emergencies call centers or travel companies for example, the clients are to be answered within a very limited time. The Quality of Service is of prime importance in the management of call centers. In addition, the staff agents cost in call centers represents 60 % to 80 % of the total operating budget [1]. Thus firms have to propose a satisfying service while controling the cost of the manpower. The importance of this sector in the service economy and the practical inherent constraints of the scheduling problem make this problem a topical issue in Operations Research.

Practically, scheduling call centers consists in deciding how many agents handling the phone calls should be assigned to work in the forthcoming days or weeks. The goal is to minimize the manpower cost while respecting a chosen

© Springer International Publishing Switzerland 2015
D. de Werra et al. (Eds.): ICORES 2015, CCIS 577, pp. 3–20, 2015.
DOI: 10.1007/978-3-319-27680-9_1

Quality of Service (QoS). In call centers, we usually consider the expected waiting time before being served, or the expected number of clients hanging up before being served, i.e. the abandonment rate, as a relevant measure of Quality of Service.

The standard model for this problem is based on forecasts of expected call arrival rates. These forecasts are computed from historical data giving the numbers of calls for the working time horizon. Since the quantity of calls vary strongly in time, the working horizon is split in small periods of time, usually 30-minute periods. Thus we obtain for each period an expected call arrival rate. Then we are able to compute the staff requirements for each period from the forecasts and an objective service level which represents the chosen Quality of Service. This computation is done with the well-known Erlang C model. Finally, the numbers of agents required for the whole working horizon are determined through an optimization program, using the previous period-by-period results.

The shift-scheduling problem presents some characteristics: first, we need to split the horizon into small periods of time in order to be able to represent the variation of rate with the best precision possible. This leads to an increasing number of variables. Second, since we are considering human agents we have to respect several manpower constraints. Thus, agents have to follow established shifts and can not work only for a few hours. Moreover, the solution of the problem represents humans, so it has to be integers. Finally, call arrival rates are forecasts and thus subject to uncertainty. Thus, the final numbers of agents computed is subject to uncertainty as well. This should be considered in order to propose a valid model.

Typical call centers models consider a queuing system for which the arrival process is Poisson with known mean arrival rates [6]. Since the data of the problem are forecasts of arrival rates, the accuracy of this deterministic approach is limited. Indeed, these estimations of mean arrival rates may differ from the reality. Uncertainty is taken into account in several papers, with various approaches. Several published works consider that input parameters of the optimization program follow known distributions. Some deal with continuous distributions [5], discrete distributions [12] or discretizations of a continuous distribution into several possible scenarios [11,13] or [7]. However it can be difficult to estimate which distribution is appropriate. [10] for call centers and [4] for general problems consider a distributionally robust approach. The problem deals with minimizing the final cost considering the most unfavorable distribution of a family of distributions whose parameters are the given mean and variance. In [10], the χ^2 statistic is used to build the class of possibles discrete distributions, with a confidence set around the estimated values. [4] consider the set of radial distributions to characterise the uncertainty region, but do not solve the final optimization program for this set. Moreover they do not focus on a specific problem and do not consider integer variables.

In the optimization program, we need to take into account and manage the risk of not respecting the objective service level. [11,13] choose to penalize the non respect of the objective service level with a penalty cost in the objective function of the optimization program. [5,9] use a chance-constrained model, in which the constraints are probabilities to be respected with the given risk level. [9] focus on the staffing problem but not the scheduling problem, and consider only one period of time.

The contributions of this paper are the following: first we model our problem with uncertain mean arrival rates and a joint chance-constrained mixed-integer linear program. This approach corresponds well with the real requirements of the scheduling problem in call centers. Indeed, forecasts are a useful indication of what can happen in reality but can not be considered as enough. This approach is in contrast with most previous publications whose risk management rely on a penalty cost. This penality can be difficult to estimate.

Second we consider the risk level on the whole horizon of study instead of period by period with joint chance constraints. It enables to control the Quality of Service on the whole horizon of study, which is a critical benefit. Managers demand to have a weekly vision of the call center, and not only for short periods of time. Moreover we propose a flexible sharing out of the risk through the periods in order to guarantee minimization of the costs. As far as we know, this consideration is only used in [5] for the staffing and scheduling problem in call centers.

Finally we focus on a distributionally robust approach, considering that we only know the first two moments of the continuous probability distributions. Since we do not know in reality what is the adequate distribution, we investigate a way of solving the problem for unknown distributions. Unlike other proposed distributionally robust approaches ([10] in particular), we consider continuous distributions instead of discrete distributions. This allows to a better representation of the reality. Moreover, [10] focus on the uncertainty on the parameters of a known gamma distribution whereas we focus on the uncertainty of the distribution with known parameters.

The rest of the paper is organized as follows. In Sect. 2 we present the formulation of the problem. At first, we propose the staffing model used for computing the useful data of the scheduling problem. Then we introduce the distributionally robust chance-constrained approach. In Sect. 3 we propose computations leading to the deterministic equivalent of the distributionally robust program. We also present the piecewise linear approximations leading to the final programs whose solutions are lower and upper bounds of the initial optimal solution. Section 4 gives some numerical results. Finally 5 investigates the importance of the choice of the distribution and thus the benefit of the distributionally robust approach.

2 Problem Formulation

2.1 Staffing Model

The shift-scheduling problem is induced by the fact that we consider whole number of human agents working according to manpower constraints. We have to consider that agents can not come and work for only a few hours and need to follow working shifts of full-time or part-time jobs. These shifts are made up of working hours and breaks, for lunch for example. The problem is then to decide how many working agents need to be assigned to each shift in the call center in order to respect a choosen objective service level. This computation uses data of calls arrival rates.

As previously explained, since arrival rates vary strongly in time, the horizon is split into T small periods, typically 15 or 30 min. For each small period of time t, forecasts are computed from historical data of numbers of calls. Based on these forecasts of number of incoming calls, we can compute the agents requirements at each period of time t.

In that goal we use the Erlang C model, [6]. At each period of time t we consider the call center as a queuing system in stationary state [8]. This is a $M_t/M/N_t$ queue, where the customer arrival process is Poisson with rate λ and the services times are independent and exponentially distributed with rate μ. The number of servers, i.e. number of agents of our problem, is denoted by N_t for the period t. The queue is assumed to have an infinite capacity, with a First Come-First Served (FCFS) discipline of service.

In our problem we consider the average waiting time as the Quality of Service. The Erlang C model gives the function of Average Speed of Answer (ASA). This function gives the expected waiting time according to the parameters of the queue: the service rate μ, the arrival rate λ and the number of servers N. The ASA function is the following (see [6] or [3]):

$$ASA(N, \lambda, \mu) = \mathbb{E}[Wait] \tag{1}$$
$$= P\{Wait > 0\}\mathbb{E}[Wait|Wait > 0] \tag{2}$$
$$= \frac{1}{N * \mu * (1 - \frac{\lambda}{N*\mu}) \left(1 + (1 - \frac{\lambda}{N*\mu}) \sum_{m=0}^{N-1} \frac{N!}{m!}(\frac{\mu}{\lambda})^{N-m}\right)}$$

Note. In this relation λ and μ are real numbers whereas N is an integer. In the studied problem, the objective service level is a maximum ASA value. We denote ASA^* this value. As in [5], we will introduce a function of λ, μ and ASA^* giving the required number of agents, which will be here considered as a real value.

2.2   Computation of Staffing Requirements

The previous ASA (Average Speed of Answer) function is used in an algorithm to compute the minimum number of agents required to reach the targeted ASA^*, given λ and μ.

The procedure is the following:

– We compute $ASA(N, \lambda, \mu)$ and $ASA(N + 1, \lambda, \mu)$ such that

$$ASA(N, \lambda, \mu) \geqslant ASA^* \text{ and } ASA(N + 1, \lambda, \mu) < ASA^*$$

We denote $ASA(N, \lambda, \mu)$ as $ASA_{N,\lambda}$.

– The real value of N is computed by a linearization in the $[ASA_{N,\lambda}; ASA_{N+1,\lambda}]$ segment. The affine function is:

$$ASA^* = (ASA_{N+1,\lambda} - ASA_{N,\lambda}) * b$$
$$+ (N+1) * ASA_{N,\lambda} - N * ASA_{N+1,\lambda}$$

and b is the real value of required agents we are looking for. $\square$

For each period, this algorithm gives us the requirement value b as a function of λ.

$$b = \frac{ASA^* + N * ASA_{N+1,\lambda} - (N+1) * ASA_{N,\lambda}}{ASA_{N+1,\lambda} - ASA_{N,\lambda}} \tag{3}$$

Note. For a simpler reading we chose the $ASA_{N,\lambda}$ notation instead of $ASA_{N,\lambda,\mu}$. Finally we are able to compute the number of agents b required to respect the objective service level ASA^* when the clients arrive at the rate λ and they are served at the rate μ.

The values of b obtained represent estimations of agents requirements. Since our computed results are subject to uncertainty, we consider that they are in fact the means of random variables of requirements. By considering real values rather than integers through the previous algorithm, we ensure a better precision in the uncertainty management. We assume that these variables are independent.

In next section, we present the distributionally robust optimization program for solving the shift-scheduling problem, considering the agents numbers as random variables.

2.3 Distributionally Robust Model

We consider the following chance-constrained shift-scheduling problem:

$$\min c^t x \tag{4}$$
$$\text{s.t. } P\{Ax \geqslant b\} \geqslant 1 - \epsilon$$
$$x \in (\mathbb{Z}^+)^S, \epsilon \in]0;1]$$

where c is the cost vector, x is the agents vector, b is the vector of agents requirements b_t and A is the shifts matrix. The matrix $A = (a_{i,j})_{[1;T] \times [1;S]}$ is the matrix of S shifts of T periods. The term $a_{i,j}$ is equal to 1 if agents are working during period i according to shift j and 0 otherwise. The agents vector x is composed of S variables; x_i is the number of agents assigned to the shift i. Thus there are T constraints, each for one period of time, and the product Ax represents the number of assigned agents for each of these periods. Finally, ϵ is the risk we allow us to take. Then $1 - \epsilon$ is the confidence interval.

This program minimizes the manpower cost of working agents while respecting the chosen objective service level for the horizon time under the risk level ϵ. The objective service level is the value ASA^* described in previous section. Thus we want to guarantee a maximum expected waiting time for the client while controlling the costs.

The chance constraints approach is chosen in order to deal with random variables. We want to guarantee that the probability that we staffed enough agents is higher than the given proportion $1 - \epsilon$. Then, our program deals with joint chance constraints. Indeed, instead of considering individual constraints and one risk level for each period, we set the risk for the whole horizon time.

We assume that we do not know exactly what distributions the random variables b_t are following, but we know the means $\bar{b}_t$ and the variances σ_t^2. We focus here on the distributionally robust approach: we do not know which distribution is the correct distribution but we want to optimize our problem for all the possible distributions and thus the most unfavourable distribution with known expected value and variance. We note $b \sim (\bar{b}, \sigma^2)$ the vector of variables b_t, with means $\bar{b}_t$ and variances σ_t^2.

Then, we consider the following program:

$$\min c^t x \tag{5}$$
$$\text{s.t.} \inf_{b \sim (\bar{b}, \sigma^2)} P\{Ax \geqslant b\} \geqslant 1 - \epsilon$$
$$x \in (\mathbb{Z}^+)^S, \epsilon \in]0; 1]$$

Since we assume that the random variables are independent, we can split the constraint into T independent constraints. We propose here to dynamically share out the risk through the periods. Indeed, instead of choosing how to share out the risk through the periods before the optimization process, we decide that the proportion for each period will be a variable of the optimization program. This flexibility leads to cheaper solutions and are still satisfactory in term of robustness [5].

We introduce the variables y_t which represent the proportion of risk allocated to each period t:

$$\min c^t x \tag{6}$$
$$\text{s.t.} \ \forall t \in [\![1; T]\!], \inf_{b_t \sim (\bar{b}_t, \sigma_t^2)} P\{A_t x \geqslant b_t\} \geqslant (1 - \epsilon)^{y_t}$$
$$\sum_{t=1}^{T} y_t = 1$$
$$x \in (\mathbb{Z}^+)^S, \epsilon \in]0; 1], \ \forall t \in [\![1; T]\!], \ y_t \in]0; 1]$$

The sum of the variables y_t should be equal to 1 in order to reach the chosen risk level. In the next subsection, we give a deterministic equivalent of the chance constraints of the problem.

3   Deterministic Equivalent Problem

3.1   Dealing with the Constraints

Let us focus on the expression of one constraint. For a given period t, we have:

$$\inf_{b_t \sim (\bar{b}_t, \sigma_t^2)} P\{A_t x \geqslant b_t\} \geqslant (1 - \epsilon)^{y_t} \tag{7}$$

Using [2] (Prop.1), we obtain the following result :

$$\sup_{b_t \sim (\bar{b}_t, \sigma_t^2)} P\{A_t x < b_t\} = \begin{cases} \frac{\sigma_t^2}{\sigma_t^2 + (A_t x - \bar{b}_t)^2} & \text{if } A_t x \geqslant \bar{b}_t \\ 1 & \text{otherwise} \end{cases} \tag{8}$$

Then, considering

$$\inf_{b_t \sim (\bar{b}_t, \sigma_t^2)} P\{A_t x \geqslant b_t\} = 1 - \sup_{b_t \sim (\bar{b}_t, \sigma_t^2)} P\{A_t x < b_t\}$$

The constraint (7) is respected if and only if

$$A_t x - \bar{b}_t \geqslant 0, \quad \frac{(A_t x - \bar{b}_t)^2}{\sigma_t^2 + (A_t x - \bar{b}_t)^2} \geqslant (1 - \epsilon)^{y_t}$$

Then we can give an equivalent to the constraint:

$$\inf_{b_t \sim (\bar{b}_t, \sigma_t^2)} P\{A_t x \geqslant b_t\} \geqslant (1 - \epsilon)^{y_t}$$

$$\Leftrightarrow \frac{(A_t x - \bar{b}_t)^2}{\sigma_t^2 + (A_t x - \bar{b}_t)^2} \geqslant (1 - \epsilon)^{y_t}$$

$$\Leftrightarrow \frac{(A_t x - \bar{b}_t)^2}{\sigma_t^2} \geqslant \frac{(1 - \epsilon)^{y_t}}{1 - (1 - \epsilon)^{y_t}}$$

We note $p = 1 - \epsilon$ and since $A_t x - \bar{b}_t \geqslant 0$, we have the result

$$\frac{A_t x - \bar{b}_t}{\sigma_t} \geqslant \sqrt{\frac{p^{y_t}}{1 - p^{y_t}}}$$

We now have a deterministic equivalent of our distributionally robust chance constraints. Finally, we propose to linearize the Right-Hand Side of the constraints and obtain bounds of the optimal solution.

3.2 Linear Approximations

We focus here on giving an upper bound and a lower bound of the problem by considering linearizations of the Right-Hand Side (RHS). Let us consider the following function, with $\epsilon \in]0; 1]$ and $p = 1 - \epsilon$:

$$f :]0; 1] \rightarrow \qquad\qquad\qquad\qquad \mathbb{R}^+ \tag{9}$$

$$y \mapsto \qquad\qquad\qquad\qquad \sqrt{\frac{p^y}{1 - p^y}}$$

By deriving this function twice, we prove that it is convex. The detail is in the Appendix. $\qquad\qquad\qquad\qquad\qquad\qquad\qquad\qquad\qquad\qquad\qquad\qquad \square$

This result guarantees that the following linearizations are upper and lower bounds of the functions, that is to say, we propose linearizations that are always above or below the function's curve.

Here is an illustration of the piecewise approximations of function f for:

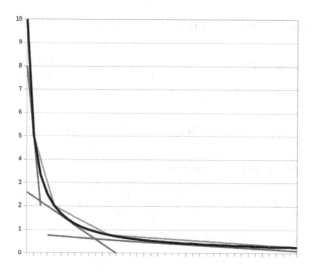

Fig. 1. Piecewise linear approximations of function f.

Piecewise Tangent Approximation. We give here a lower bound of f : $y \mapsto \sqrt{\frac{p^y}{1-p^y}}$. Since we proved the convexity of the function, we know that the piecewise tangent approximation is a lower bound.

Let us choose n points $y_j \in]0; 1]$, $j \in [\![1; n]\!]$ be n points such that $y_1 < y_2 < \ldots < y_n$.

We denote $\hat{f}_{l,j}$ the piecewise tangent approximation around the point y_j (the subscript l stands for *lower*). We compute this linearization with a first-order Taylor series expansion around the n points:

$$\forall j \in [\![1; n]\!],$$
$$\hat{f}_{l,j}(y) = f(y_j) + (y - y_j)f'(y_j) \tag{10}$$
$$= f(y_j) + (y - y_j)f(y_j)\frac{\ln p}{2(1 - p^{y_j})}$$
$$= \delta_{l,j} * y + \alpha_{l,j} \tag{11}$$

Since we proved the convexity of the function, we describe the new constraint as

$$\hat{f}_l(y) = \max_{j \in [\![1;n]\!]} \{\hat{f}_{l,j}(y)\} \tag{12}$$

In the program, this condition is expressed in each period with one constraint for each approximation point. Finally, we give a lower bound of the solution of the initial problem with the following program:

$$\min c^t x \tag{13}$$
$$\text{s.t. } \forall t \in [\![1; T]\!], \forall j \in [\![1; n]\!],$$

$$\frac{A_t x - b_t}{\sigma_t} \geqslant \delta_{l,j} y_t + \alpha_{l,j}$$

$$\sum_{t=1}^{T} y_t = 1$$

$$x \in (\mathbb{Z}^+)^S, \epsilon \in]0;1], \ \forall t \in [\![1;T]\!], \ y_t \in]0;1]$$

where S is the number of shifts and T the number of periods.

Piecewise Linear Approximation. Similarly, we give here an upper bound of the function with a piecewise linear approximation.

Let us choose n points $y_j \in]0;1]$, $j \in [\![1;n]\!]$ be n points such that $y_1 < y_2 < \ldots < y_n$ and interpolate linearly between them.

We denote $\hat{f}_{u,j}$ the piecewise linear approximation between the points y_j and y_{j+1} (the subscript u stands for *upper*):

$$\forall j \in [\![1;n-1]\!],$$
$$\hat{f}_{u,j}(y) = f(y_j)$$
$$+ \frac{y - y_j}{y_{j+1} - y_j} * (f(y_{j+1}) - f(y_j))$$
$$= \delta_{u,j} * y + \alpha_{u,j} \tag{14}$$

Again our new program respects the constraint

$$\hat{f}_u(y) = \max_{j \in [\![1;n]\!]} \{\hat{f}_{u,j}(y)\} \tag{15}$$

Finally, the following program gives an upper bound of our problem:

$$\min c^t x \tag{16}$$
$$\text{s.t. } \forall t \in [\![1;T]\!], \ \forall j \in [\![1;n-1]\!],$$
$$A_t x - b_t \geqslant (\delta_{u,j} y_t + \alpha_{u,j}) \sigma_t$$
$$\sum_{t=1}^{T} y_t = 1$$
$$x \in (\mathbb{Z}^+)^S, \epsilon \in]0;1], \ \forall t \in [\![1;T]\!], \ y_t \in]0;1]$$

where S is the number of shifts and T the number of periods.

In this section we first proposed a deterministic equivalent to the initial distributionally robust stochastic problem. Therefore, the optimal solution of the deterministic program is the optimal solution of the initial program. We had to deal with a mixed-integer nonlinear program. Second, we provided close upper and lower bounds of the optimal solution by introducing piecewise tangent and linear approximations. This was possible because of the convexity of the constraints. This led to two mixed-integer linear programs whose number of integer and binary variables are not increased compared to the initial formulation. These

Table 1. Results for different parameters.

Parameters				Results	
Variance range	λ range	ASA^*	Risk ϵ	Cost gap	% Violations
0.3–1	16–86	1	15 %	0.0045	5–3
0.3–5	16–86	1	15 %	0.011	1–0
0.1–1	4–20	1	15 %	0.027	13–9
0.1–1	4–20	1	10 %	0.034	7–4
2.5–9	4–20	0.3	15 %	0.052	1–1

two programs are easily computed with an optimization software (CPLEX for example). This enables to give bounds of the optimal solution of the initial complex problem.

Next section gives an example of the method to solve a scheduling problem in call centers.

4 Numerical Experiments

4.1 Instance

In order to evaluate the quality and the robustness of our model, we applied our approach to instances based on data from a health insurance call center. This data provides forecasts for one week from Monday morning to Saturday midday (5 days of 10 h and 1 half-day of 3.5 h). The horizon is split in 30 minute periods. 24 differents shifts, from both full-time and part-time schedules, make up the shifts matrix. As we previously said in Sect. 2, we can standardize the service rate μ without loss of generality. We consider that all agents have the same hourly salary, thus the cost of one agent is proportional to the number of periods worked.

We computed the vectors of scheduled agents x_l and x_u for one week with the two programs (13) and (16) of the previous section, providing an upper bound and a lower bound of the optimal solution cost. We used 17 points for computing the piecewise tangent and linear approximations. We noticed that the order of magnitude of variables y_t is between 10^{-2} and 10^{-1}, thus we reduced the gap between the upper and lower bounds by gathering most of the points around this area.

We want to evaluate the quality of our solutions x_l and x_u. To this end we simulate possible realizations of arrival rates according to different distributions with the same data as previously. We consider different possible distributions: gamma distributions, uniform distribution, Pareto distribution, and variations of normal distributions (log-normal, folded normal).

We elaborate a scenario as following: for each period of time we simulate a call arrival rate according to one of the given probability distributions. Then we compute the number of effective required agents for each period. A scenario covers requirements for the whole time horizon. Finally we compare these values of requirements with our solutions of the problem (lower solution x_l and upper solution x_u). A scenario is considered as violated if at least in one period the scheduled solution by x_u or x_l is not enough in comparison of what the realization requires.

We computed between 100 and 500 scenarios for each probability distributions. The percentage of violations gives us an idea of the robustness of our approach for several chosen distributions. The cost of the solutions gives us an idea of the quality of the minimization.

4.2   Results

In Table 1, we give the percentage of violated scenarios for various ranges of values of means and variances, and risk level. The queue parameters μ was set to 1 as it simply represents a multiplicity factor. The first column gives the range of values of the variances through the day. The second column gives the range of values of the means through the day, following a typical seasonality.

The value Cost Gap (CG) of the 5th column is given by the relative difference between the cost of the upper bound solution and the cost of the lower bound solution: $CG = \frac{c^t x_u - c^t x_l}{c^t x_l}$.

The last column gives the number of violated scenarios for the lower bound and for the upper bound.

In Table 1 we can notice that both upper and lower bound solutions respect the set risk level. The variations of the parameters show that the bigger the variances, the better the model. The distributionally robust model deals very well with increasing of variances. We notice that even if we allow 15 % risk, only a few scenarios are violated when the variances are higher (second and last lines of Table 1). In these cases the call center is over-staffed and the given solutions seem too conservative. But it is important to remember that all the observations are based on simulations of only a few examples of distributions. These very low percentages only show that if the arrival rates λs follow in reality one of the studied distributions, it may be over-staffed. However the distributionally robust model indeed consists in taking all possible distributions with given mean and variance into account. Thus it may be possible to reach the maximum risk level with other particular distributions.

These results show that our approach is robust, considering the numbers of violations never exceed the risk level we set. The values of Cost Gap show that the two bounds are close enough to propose a very close solution to optimal solution.

We can notice that even if the solutions costs are very close, the number of violations is different between the upper solution and the lower solution. This is due to the fact that the distribution of the agents through the different shifts is different according to the programs.

Table 2 focuses on comparing results for different risk levels. The simulations were made with these parameters:

Table 2. Results for different risk levels.

Parameters		Results				
ASA^*	Risk ϵ	Lower cost	Upper cost	Cost Gap	% Violations upper solution	% Violations lower solution
5	15 %	27	27.75	0.028	5	3
5	10 %	29	30	0.034	2	1
5	05 %	33.81	35.31	0.044	0	0
1	15 %	27.5	28.5	0.036	7	4
1	10 %	29.44	30.5	0.036	3	2
1	05 %	34.31	35.88	0.046	1	1
0.3	15 %	28.88	29.94	0.037	12	7
0.3	10 %	31.12	32.19	0.034	5	2
0.3	05 %	35.69	37.56	0.052	0	0

- $\mu = 1$
- λ follows a daily seasonality, varying between 4 calls/min and 21 calls/min
- σ^2 varies through the periods, between 0.25 and 1.

These parameters show well the performance of the model. Table 2 gives the costs of the two bound solutions and the Cost Gap. Like previously we ran 100 simulations and evaluated the number of violated scenarios, which is given in the last column of the table.

The first two columns of Table 2 gives the chosen parameters. Columns 3 and 4 gives the solution costs of the two programs and column 5 gives the Cost Gap. Finally, the two last columns give the number of violated scenarios for the two solutions.

Unsurprisingly, the cost of the solution increases when the risk level decreases. The Cost Gap seems to remain in a small range, even if we notice a small increase of the gap when the risk is lowered.

We can also see an increasing of the cost when ASA^* (the objective Average Speed of Answer) decreases.

Like previously, the violations results show that our model respects the initial risk conditions, for both upper and lower solutions.

Figure 2 show the values of y_t variables through the horizon for the upper bound (in blue) and the lower bound (in green). The red line shows the equal division of the risk through the day. This figure brings out the interest of dynamically sharing out the risk: optimization of the variable y_t shows their value are different from the simple equal division through the periods. Thus our approach is more complicated but leads to cheaper solutions than a simpler approach with fixed risk levels.

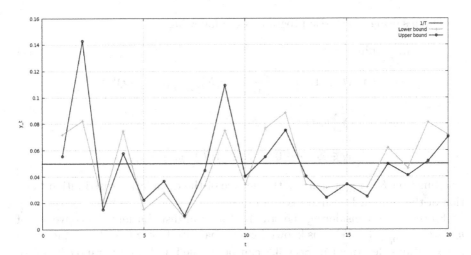

Fig. 2. Sharing out of the risk through the day.

5   Stochastic Approach with a Wrong Assumption vs Distributionally Robust Approach

5.1   Stochastic Program with Normal Distribution

In the previous section we highlighted the robustness of the distributionally robust approach. We also noticed that the solution computed by our approach may be overstaffed. However this overstaffing may be understandable since the main consideration is that we do not know what is the right distribution. In this section, we propose to compare our approach with a standard stochastic approach under the assumption that we know the distribution of the random variables. What if we model the problem under the assumption of a known distribution but appears to be the wrong one?

In the following programs, we suppose that the agents requirements are random variables following a known continuous distribution. We consider here the normal distribution with the known means and variances, as previously.

We derive the first stochastic program (4) introduced in Sect. 2.3 as in [5], without considering the infimum on the chance constraint.

However, the same considerations as previously are still valid: the variables are independent, we consider a dynamic sharing out of the risk and do piecewise linearizations leading to upper and lower bounds.

For easier computation, we standardize the normal distribution and denote β the standard normal deviate.

The resulting stochastic program is the following:

$$\min c^t x \qquad (17)$$

$$\text{s.t. } \forall t \in [\![1;T]\!], \quad \frac{A_t x - \bar{b}_t}{\sigma_t} \geqslant F^{-1}_{\beta \sim \mathcal{N}(0,1)} \left((1 - \epsilon)^{y_t}\right)$$

$$\sum_{t=1}^{T} y_t = 1$$

$$x_i \in \mathbb{Z}^+, \epsilon \in]0;1], \forall t \in [\![1;T]\!], \ y_t \in]0;1]$$

The function $F^{-1}_{\beta \sim \mathcal{N}(0,1)}$ denotes the inverse of the cumulative distribution function of the standard normal distribution.

Since we are considering the standard normal distribution, the convexity of $y_t \mapsto F^{-1}_{\beta \sim \mathcal{N}(0,1)} \left((1 - \epsilon)^{y_t}\right)$ is known for a probability $1 - \epsilon \geqslant 0.5$. Hence, piecewise linearizations described in Sect. 3.2 can be applied and are guaranteed to give upper and lower bounds.

The lower bound is given by a first-order Taylor series expansion on n given points for the tangent approximation. The resulting program is:

$$\min c^t x$$

$$\text{s.t.} \forall t \in [\![1;T]\!], \forall j \in [\![1;n]\!], \quad \frac{A_t x - \bar{b}_t}{\sigma_t} \geqslant \delta_j * y_t + \alpha_j \qquad (18)$$

$$\sum_{t=1}^{T} y_t = 1$$

$$\forall i \in [\![1;S]\!], x_i \in \mathbb{Z}^+, \forall t \in [\![1;T]\!], y_t \in]0;1]$$

The coefficients for the lower bound program are

$$\forall j \in [\![1;n]\!], \hat{F}^{-1}_{l,\beta j}(p^y) = F^{-1}_\beta(p^{y_j}) + (y - y_j) * (F^{-1}_\beta)'(p^{y_j}) * p^{y_j} * \ln(p) = \delta_j * y + \alpha_j$$

$$\text{where} (F^{-1}_\beta)'(p^{y_j}) = \frac{1}{F'_\beta(F^{-1}_\beta)(p^{y_j})} = \frac{1}{f_\beta(F^{-1}_\beta(p^{y_j}))}$$

Similarly, the upper bound program with n points for the linear approximation is:

$$\min c^t x \qquad (19)$$

$$\text{s.t. } \forall t \in [\![1;T]\!], \forall j \in [\![1;n-1]\!], \quad \frac{A_t x - \bar{b}_t}{\sigma_t} \geqslant \delta_j * y_t + \alpha_j$$

$$\sum_{t=1}^{T} y_t = 1$$

$$\forall i \in [\![1;S]\!], x_i \in \mathbb{Z}^+, \forall t \in [\![1;T]\!], y_t \in]\alpha_1;1]$$

The coefficients for the upper bound program are

$$\forall j \in [1; n-1], \ \hat{F}_{u,\beta_j}^{-1}(p^y) = F_\beta^{-1}(p^{y_j}) + \frac{y - y_j}{y_{j+1} - y_j} * (F_\beta^{-1}(p^{y_{j+1}}) - F_\beta^{-1}(p^{y_j}))$$

$$= \delta_j * y + \alpha_j \tag{20}$$

We now have upper bound and lower bound mixed-integer linear programs. Since we assumed that random variables are following normal distributions, we would like to evaluate the significance of this assumption, in the case it is wrong.

5.2 Comparison on Instances

Using the following parameters, we compute a lower solution x_l and an upper solution x_u of the problem:

- $ASA^* = 1$
- $\epsilon = 0.05$
- λ range: $6 - 16$ calls/min

As shown in [5], this approach is satisfactory in terms of robustness when considering that the assumption of normal distribution is right. We now would like to check the robustness of the solutions in the event of the right distribution being not the normal distribution, but another one. Indeed, our distributionally robust approach is favourable if the choice of the distribution in advance has an significant influence on the quality of the solution.

As with the previous Sect. 4, we generated scenarios based on various distributions: gamma distributions, uniform distribution, Pareto distribution, and variations of normal distributions (log-normal, folded normal). For each distribution, we computed 100 scenarios and indicated the percentage of violations of our solutions. These results are seen in Table 3.

Table 3. Robustness of the stochastic approach for different wrong distributions. Targeted risk level is **5 %**.

Distribution	Percentage of violations	
	For lower bound	For upper bound
Normal	6 %	3 %
Gamma	**15 %**	**7 %**
Folded normal	5 %	2 %
Log-normal	**10 %**	**6 %**
Pareto	3 %	1 %

Table 3 represent the percentages of violations of our solutions if the real distributions are the ones indicated. The first line shows the quality of the stochatic approach when the assumption of normal distribution is right.

The second line of the table shows that if the right distribution is the gamma distribution, both the lower bound and the upper bound staffing solutions do not give a satisfactory risk management. The solutions were computed in order to respect a risk of 5 % but 7 % to 15 % of scenarios were violated.

Thus we can notice that if we wrongly choose the normal distribution instead of the Gamma or the Log-normal distribution, the quality of the solution is not satisfactory anymore. The other distributions still respects the targeted risk level on this batch of scenarios.

Note. We noticed some rare batches where the scenarios generated with the Pareto distribution did not respect the risk level for both upper and lower bounds solutions. As far as we tested, the Folded normal distribution scenarios showed violations only for upper bound, but the percentage was high compared to the allowed risk level (between 3 and 4 times).

This result shows that if the choice of the distribution is wrong, the resulted staffing solutions can not longer be satisfactory. This problem does not appear in our distributionally robust approach, since it is designed intrinsically to deal with this difficulty. Since in some situations it is difficult to guarantee the right distribution, the distributionally robust approach is definitely adapted.

6   Conclusion

This paper presents a distributionally robust approach for the staffing and shift-scheduling problem arising in call center. We introduced the distributionally robust approach, considering that the call arrival rates are following unknown continuous distributions. Moreover, instead of considering the risk level on a period-by-period basis, we decided to set this risk level for the whole horizon of study and thus consider a joint chance-constrained program. Then, we proposed a deterministic equivalent of the distributionally robust approach with a dynamic sharing out of the risk. We were thus able to propose solutions with reduced costs compared to other published approaches. Finally we gave lower and upper bound of the problem with piecewise linear approximations. Computational results show that both upper and lower solutions respect the objective risk level for a given set of continuous distributions. This shows that our approach proposes robust solutions. The Cost Gap was small enough to be able to bring out a valid solution for the initial problem, which is eventually useful for the managers.

In the simulations, we noticed that mainly the Pareto distribution and Gamma distribution are the ones with violated scenarios. The solutions of the model show that for other distributions, the call center may be over-staffed. Thus, we could study further the call center model in order to evaluate what are the interesting distributions to consider. This can lead, as an improvment for our work in the future, to the study of a given set of distributions, according to some conditions (in addition to the known mean and variance).

The distributionally robust approach showed an advantage compared to a stochastic program with a wrong assumption on the distribution. Indeed, if the assumption of normal distribution turns out to be incorrect, the staffing solutions are not satisfactory whereas the distributionally robust approach considers this possibility per se.

Moreover, we can focus on improving the queuing system model by considering another approach of the representation of the service level in order to have a closer representation to reality.

Another interesting future research would be to conduct a sensitivity analysis that accounts for the forecast bias.

Finally we made the assumption that periods of the day are independent. In reality, we can notice a daily correlation of the periods in a call center: busy periods may appear in an entire busy day and rarely alone. Conversely, light periods should lead to an entire light day. We can then consider that the effective arrival rates depend on a *busyness* factor, which represents this level of occupation of the day.

Acknowledgements. This research is funded by the French organism DIGITEO.

Appendix

Here we give the proof of the convexity of

$$f :]0; 1] \to \mathbb{R}^+$$

$$y \mapsto \sqrt{\frac{p^y}{1 - p^y}} \tag{21}$$

with $p \in [0; 1[$.

Function f is C^∞, so we can compute the second derivative of function f. We have first:

$$\frac{df}{dy} = \frac{\frac{\ln p}{2} p^{\frac{y}{2}}(1 - p^y)^{\frac{1}{2}} + \frac{\ln p}{2} p^y (1 - p^y)^{-\frac{1}{2}} p^{\frac{y}{2}}}{1 - p^y}$$

$$= \frac{\ln(p)(1 - p^y)^{-\frac{1}{2}} (p^{\frac{y}{2}}(1 - p^y) + p^{\frac{3}{2}y})}{2(1 - p^y)}$$

$$= f(y) \frac{\ln p}{2(1 - p^y)}$$

Then,

$$\frac{d^2 f}{dy^2} = \frac{\ln p}{2} \frac{f'(y)(1 - p^y) + \ln(p)p^y f(y)}{(1 - p^y)^2}$$

$$= \frac{\ln^2(p)(1 + 2p^y)}{4(1 - p^y)^2} f(y)$$

$$= \frac{\ln^2(p)(1 + 2p^y)}{4(1 - p^y)^2} \frac{p^{\frac{y}{2}}}{(1 - p)^{\frac{y}{2}}} \tag{22}$$

Since every term of the second derivative is positive, we conclude that $\frac{d^2 f}{dy^2}$ is positive and then, f is convex.   □

References

1. Aksin, Z., Armony, M., Mehrotra, V.: The modern call center: a multi-disciplinary perspective on operations management research. Prod. Oper. Manage. **16**, 665–688 (2007)
2. Bertsimas, D., Popescu, I.: Optimal inequalities in probability theory: a convex optimization approach. Technical report, Department of Mathematics and Operations Research, Massachusetts Institute of Technology, Cambridge, Massachusetts (1998)
3. Brown, L., Gans, N., Mandelbaum, A., Sakov, A., Shen, H., Zeltyn, S., Zhao, L.: Statistical analysis of a telephone call center: a queueing-science perspective. J. Am. Stat. Assoc. **100**, 36–50 (2005)
4. Calafiore, G.C., El Ghaoui, L.: On distributionally robust chance-constrained linear programs. J. Optim. Theory Appl. **130**, 1–22 (2006)
5. Singh, T.P., Neagu, N., Quattrone, M., Briet, P.: A decomposition approach to solve large-scale network design problems in cylinder gas distribution. In: Pinson, E., Valente, F., Vitoriano, B. (eds.) ICORES 2014. CCIS, vol. 509, pp. 265–284. Springer, Heidelberg (2015)
6. Gans, N., Koole, G., Mandelbaum, A.: Telephone call centers: tutorial, review, and research prospects. Manuf. Serv. Oper. Manage. **5**, 79–141 (2003)
7. Gans, N., Shen, H., Zhou, Y.P.: Parametric stochastic programming models for call-center workforce scheduling, working paper (April 2012)
8. Gross, D., Shortle, J.F., Thompson, J.M., Harris, C.M.: Fundamentals of Queueing Theory. Wiley Series, New York (2008)
9. Gurvich, I., Luedtke, J., Tezcan, T.: Staffing call centers with uncertain demand forecasts: a chance-constrained optimization approach. Manage. Sci. **56**, 1093–1115 (2010)
10. Liao, S., van Delft, C., Vial, J.P.: Distributionally robust workforce scheduling in call centers with uncertain arrival rates. Optim. Methods Softw. **28**, 501–522 (2013)
11. Liao, S., Koole, G., van Delft, C., Jouini, O.: Staffing a call center with uncertain non-stationary arrival rate and flexibility. OR Spectr. **34**, 691–721 (2012)
12. Luedtke, J., Ahmed, S., Nemhauser, G.L.: An integer programming approach for linear programs with probabilistic constraints. In: Fischetti, M., Williamson, D.P. (eds.) IPCO 2007. LNCS, vol. 4513, pp. 410–423. Springer, Heidelberg (2007)
13. Robbins, T.R., Harrison, T.P.: A stochastic programming model for scheduling call centers with global service level agreements. Eur. J. Oper. Res. **207**, 1608–1619 (2010)

A Comparison of a Global Approach and a Decomposition Method for Frequency Assignment in Multibeam Satellite Systems

Jean-Thomas Camino[1,2,3]([✉]), Christian Artigues[2,3], Laurent Houssin[2,4], and Stéphane Mourgues[1]

[1] Telecommunication Systems Department, Airbus Defence and Space, Space Systems, 31 Rue des Cosmonautes, 31402 Toulouse, France
{jean-thomas.camino,stephane.mourgues}@astrium.eads.net
[2] CNRS, LAAS, 7 Avenue du Colonel Roche, 31400 Toulouse, France
{artigues,houssin}@laas.fr
[3] Université de Toulouse, LAAS, 31400 Toulouse, France
[4] Université de Toulouse, UPS, LAAS, 31400 Toulouse, France

Abstract. As a result of the continually growing demand for multimedia content and higher throughputs in wireless communication systems, the telecommunication industry has to keep improving the use of the bandwidth resources. This access to the radiofrequency spectrum is both limited and expensive, which has naturally lead to the definition of the generic class of combinatorial optimization problems known as "Frequency Assignment Problems" (FAP). In this article, we present a new extension of these problems to the case of satellite systems that use a multibeam coverage. With the models we propose, we make sure that for each frequency plan produced there exists a corresponding satellite payload architecture that is cost-efficient and decently complex. Two approaches are presented and compared: a global constraint program that handles all the constraints simultaneously, and a decomposition method that involves both constraint programming and integer linear programming. For the latter approach where two subproblems are studied, we show that one of them can be modeled as a multiprocessor scheduling problem while the other can either be seen as a path-covering problem or a multidimensionnal bin-packing problem depending on the assumptions made. These analogies are used to prove that both the subproblems addressed in the decomposition method belong to the category of NP-hard problems. We also show that, for the most common class of interference graphs in multibeam satellite systems, the maximal cliques can all be enumerated in polynomial time and their number is relatively low, therefore it is perfectly acceptable to rely on them in the scheduling model that we derived. Our experiments on realistic scenarios show that the decomposition method proposed can indeed provide a solution of the problem when the global CP model does not.

Keywords: Frequency assignment · Multiprocessor scheduling · Path cover · Linear programming · Constraint programming · Maximal cliques enumeration

© Springer International Publishing Switzerland 2015
D. de Werra et al. (Eds.): ICORES 2015, CCIS 577, pp. 21–39, 2015.
DOI: 10.1007/978-3-319-27680-9_2

1   Introduction

A common characteristic of any telecommunication system is that it is band-width limited, and one of the main challenges for the system engineers is to optimally use this precious resource. Satellite telecommunications systems are no exception to that rule, and this already difficult task is even more complex when the specific limitations and needs of the satellite payload are taken into consideration. Plenty of literature can be found on the problem of assigning frequencies under the name of "Frequency Assignment Problems" (FAP). For instance, [1] is a very thorough survey on the models and the optimization methods that have been developed over the years to solve the frequency assignment problems that emerged in a lot of different wireless communications systems. The recent literature proposes more and more sophisticated methods to solve the FAP, such as parallel hyperheuristics [12], differential evolution [10], population-based heuristics [8,17] or considers more and more realistic variants of the FAP according to specific problem characteristics [7,9,16]. This article aims at presenting new models and approaches for this extension of the frequency assignment problem to multibeam satellite systems, and promising results on realistic scenarios.

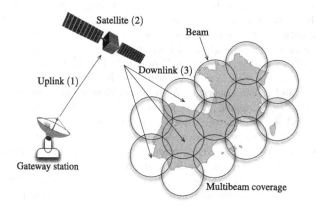

Fig. 1. The uplink (1), the satellite payload (2) and the downlink (3) of the forward link of a multibeam satellite system.

A multibeam satellite system is characterized by a plurality of relatively narrow beams used to provide coverage to its service area as shown in Fig.1, each beam being the representation of an antenna gain loss threshold for the corresponding satellite radio source. Still in Fig.1, the role of the satellite payload (2) is to receive, downconvert, amplify, and retransmit the signals of the uplink (1) in the different beams of the downlink (3) where the end-users are located. It is assumed that the system bandwidth is divided into identical frequency channels, the bandwidth of a channel being equal to that of one carrier signal. For each beam, it is either specified by the operator or assessed in advance how much bandwidth is needed and therefore how many carriers must be transmitted in it. Assuming that the carrier uplink frequencies are known or treated afterwards, system engineers have to define for each carrier of each beam:

⋄ The frequency channel used in the downlink
⋄ The polarization of the signal in the downlink
⋄ The high power amplifier in the payload that will be amplifying the corresponding uplink carrier

These are the variables of the problem presented in this paper. Values must be assigned to them with the goal to minimize the levels of interferences in each beam, the number of high power amplifiers needed in the satellite payload, and the number of hardware needed for the downconversions. More precisely, the approach we have selected is to aim at minimizing the number of high power amplifiers needed in the satellite payload since they are heavy, expensive, and highly power-consuming, while we will be using constraints to limit the interferences and the hardware needed for the downconversions to what is acceptable.

The rest of the article is structured as follows. In Sect. 2, the problem constraints are listed and detailed. Then, Sect. 3 focuses on the different approaches we have devised to actually model the problem. Finally, Sect. 4 provides experimental results and concrete scenario examples, before some concluding remarks in Sect. 5.

2   The Problem Constraints

2.1   Frequency Related Constraints

For the quality of transmission of a signal, the interferences are a determining factor and any frequency assignment procedure should try to minimize them. Let us remind that a frequency and a polarization must be assigned to each carrier of each beam in the downlink. Note that in this work, the isolation of the signals through the time-dimension is not considered. In the end, the frequency related constraints that are taken into account here are the following:

– **Polarization Isolation**
  A perfect radio antenna transmits and receives waves in a particular polarization and is insensitive to orthogonally polarized signals [4], meaning that the same frequency channel can therefore be used twice in the same area without risking severe interferences. In actual facts, antennas cannot transmit and receive perfectly in one polarization only, it is always a combination of two orthogonal polarizations, one of them being predominant. To take advantage of that property anyway, the choice here has been to consider that two carriers at the same frequency using orthogonal polarizations are allowed to be transmitted in closer zones than two carriers transmitted at the same frequency and with the same polarization.

- **Spatial Isolation**
  Thanks to antenna gain losses, two carriers can use the same color (frequency or frequency-polarization couple) as long as the two corresponding beams are sufficiently distant from each other. This is often turned into a constraint of minimum distance between them, leading the very classic binary interference constraints. The resulting representation is a graph $G = (B, E)$ where each vertex $b \in B$ corresponds to the zone covered by a beam and each edge $e \in E$ is a link between two zones where it is not allowed to use the same color.
- **Limit on the Frequency Channel Reuse Values**
  Defining an upper-bound for these values allows to balance the number of times each channel is used, which reduces the hardware needs for frequency conversions. Since two uplink carriers can only share a downconverter in the satellite payload if they need the same frequency downconversion, it is interesting to be able to define the uplink frequencies so as to have as many of these situations as possible, and this balance of the frequency reuse factors in the downlink is advantageous on that regard.

2.2   Amplification of the Signals Constraints

A traveling-wave tube (TWT) is a type of high power amplifier for radio frequency signals and a widely used technology for satellite telecommunication payloads [4]. A TWT must be assigned to each carrier of each beam under the following constraints:

- **Minimization of the Number of TWT**
  A TWT is an expensive technology, one should therefore aim at finding a distribution of the carriers in the TWTs that minimizes their number.
- **Frequency Ranges**
  The TWTs can have a bandwidth narrower than the overall system bandwidth. In that case, payload engineers agree with the equipment manufacturer on a limited number of frequency ranges. Therefore, the assignment of carriers to the TWTs must guarantee that the frequency ranges are supported by the available equipment.
- **Carriers Forbidden to Use the Same TWT**
  Two carriers cannot be amplified by the same TWT if their amplification requirements are too different, because of the non-linearity of the TWT. These incompatibilities are known in advance.
- **Single use of the Frequency Channels**
  A TWT cannot amplify two carriers using the same frequency channel.
- **Limited Number of Carriers per TWT**
  A TWT is characterized by its output power level. That power is shared by the carriers, therefore the number of carriers per TWT is upper-bounded.
- **Contiguity of the Frequencies**
  The payload complexity can be significantly reduced when there are no frequency gaps between the carriers in the same TWT. The satisfaction of this particular constraint is not systematically required, and this explains the two different models we proposed - with and without this constraint - for the traveling-wave tube assignment problem described further in the article.

3   Models

The first model we derived is a global constraint program (Sect. 3.1) that includes all the aforementioned constraints. It has been able to provide really interesting system solutions on some scenarios, however, when the number of variables is set to high realistic values, the global CP model fails at providing solutions or proving unfeasibility in reasonable time. That is why a decomposition method has been developed, with a subdivision of the problem into a multiprocessor scheduling (Sect. 3.2) and a path-covering (Sect. 3.3) problems. The two approaches, the single constraint programming model and the combination of the two submodels, are then compared experimentally in Sect. 4.

3.1   Global Constraint Programming Model

The idea to derive a constraint programming model has been motivated by an analysis of the constraints on the problem variables (frequency, polarization, TWT) that revealed that global constraints could be used to model a large part of the problem. A global constraint [2] is a set of constraints for which it is preferable to treat that set of constraints as a whole than to treat all the constraints of that conjunction of constraints individually. Using global constraints is a way to have a better view on the structure of the problem, which is then exploited with powerful filtering algorithms. On that regard, a very significant example is the all different constraint [15]

$$\text{alldifferent}(X)$$

that forces all the variables of the array X to be different. In the model below, we also use the global cardinality constraint

$$\text{global\_cardinality\_constr}(X, Y, m, M)$$

that allows to bound the number of times some items appear in a list, X being that list, Y the set of sought values, m the array of minimum number of occurrences for each sought value, M the array of maximum number of occurrences for each sought value. Finally, the Gecode convexity global constraint

$$\text{convex}(X)$$

is used to force the integers of an integer set X to be a convex sequence ($\{1, 2, 3\}$ is one while $\{1, 2, 4\}$ is not). These global constraints are implemented in the open source solver Gecode [11] that we chose to use.

An instance of this particular frequency assignment problem is defined by a set of N_B beams, each beam $b \in B = \{1, \cdots, N_B\}$ being characterized by the number n_b of carriers transmitted in it, leading to an overall number of carriers

$$N_C = \sum_{b=1}^{N_B} n_b$$

For all $b \in B$ and for all $c \in \{1, \cdots, n_b\}$,

$$\mathrm{ind}(b, c) = c + \sum_{\tilde{b}=1}^{b-1} n_{\tilde{b}}$$

defines a 1D sorting of these carriers and for all $b \in B$,

$$C_b = \{\mathrm{ind}(b, c) \mid c \in \{1, \cdots, n_b\}\}$$

is the notation for the set of indices of the carriers of the bth beam. Therefore, note that the C_b sets partition the set $C = \{1, \cdots, N_C\}$. The system bandwidth is divided into N_F sub-channels indexed by $F = \{1, \cdots, N_F\}$. N_T TWTs are available in the payload, and N_P orthogonal polarizations are considered (typically $N_P = 2$), the corresponding index sets being respectively denoted by T and P. Each carrier $c \in C$ must be assigned a frequency channel $\mathbf{f}_c \in F$, a TWT $\mathbf{t}_c \in T$ and a polarization $\mathbf{p}_c \in P$. These are the problem variables. Two graphs $G = (B, E)$ and $G' = (B, E')$ with $E' \subset E$ are defined: an edge of E' forbids the carriers in the two corresponding beams to use the same frequency channel whatever the polarization, whereas an edge of E only forbids the multiple use of the same frequency-polarization couple. In the following equations, note that $\mathrm{card}(X)$ denotes the cardinality of the set X. Here follows the list of the constraints expressed with these variables:

– For a given beam b such that $n_b > 1$, the n_b carriers must be contiguous in frequency, use the same TWT, and have the same polarization. For such b values, the constraints are:

$$\forall i \in \{2, \cdots, n_b\}, \quad \mathbf{t}_{\mathrm{ind}(b,1)} = \mathbf{t}_{\mathrm{ind}(b,i)} \tag{1}$$

$$\mathbf{p}_{\mathrm{ind}(b,1)} = \mathbf{p}_{\mathrm{ind}(b,i)} \tag{2}$$

$$\mathbf{f}_{\mathrm{ind}(b,i-1)} = \mathbf{f}_{\mathrm{ind}(b,i)} - 1 \tag{3}$$

– As discussed in Sect. 2.1, channel reuse bounds are a tunable parameter in input used to limit hardware needs for the downconversions. Let $\mathbf{R_{min}}$ and $\mathbf{R_{max}}$ be the arrays of size N_F of these bounds (note that in practice the lower-bound array is set to 0, it is just there to fit the definition of the global constraint that use both arrays), then the corresponding is the following:

$$\mathrm{global\_cardinality\_constr}(\mathbf{f}, F, \mathbf{R_{min}}, \mathbf{R_{max}}) \tag{4}$$

– The binary interference constraints associated to E can be expressed as follows for all $b, b' \in B$ such that $b < b'$ and $(b, b') \in E$:

$$\mathrm{alldifferent}(\mathbf{f}_c + N_F(\mathbf{p}_c - 1) \mid c \in C_b \cup C_{b'}) \tag{5}$$

– And for E', for all $b, b' \in B$ such that $b < b'$ and $(b, b') \in E'$:

$$\mathrm{alldifferent}(\mathbf{f}_c \mid c \in C_b \cup C_{b'}) \tag{6}$$

– The same frequency cannot be used twice by the carriers of a given TWT:

$$\forall t \in T, \forall f \in F, \mathrm{card}(\mathbf{T}_t \cap \mathbf{F}_f) \leq 1 \qquad (7)$$

where $\mathbf{T}_t \subset C$ and $\mathbf{F}_t \subset C$ respectively are the set of carriers using the TWT t and the set of carriers using the frequency channel f, these set variables being linked to the arrays $\mathbf{t}$ and $\mathbf{f}$ by side channeling constraints that we do not provide here for the sake of conciseness.

– The contiguity in the TWTs. Let us denote by $\mathcal{F}_t$ the set of frequency channels used in the TWT t, these set variables being easily defined with channeling constraints involving the variable arrays $\mathbf{f}$ and $\mathbf{t}$. Then, the global constraint convex does exactly what is sought:

$$\forall t \in T, \mathrm{convex}(\mathcal{F}_t) \qquad (8)$$

– The maximum number of carriers in a given TWT that is upper bounded by a tunable parameter n:

$$\forall t \in T, \mathrm{card}(\mathbf{T}_t) \leq n \qquad (9)$$

– The incompatibilities between the carriers that cannot use the same TWT. Let $c, c' \in C$ be two carriers forbidden to use the same TWT, then the corresponding constraint is the following:

$$\mathbf{t}_c \neq \mathbf{t}_{c'} \qquad (10)$$

– The content of the TWTs must be of a given type. Let $F_1 \subset F$ and $F_2 \subset F$ be two subparts of the system bandwidth such that $F_1 \cup F_2 = F$. These two sets define two types of acceptable frequency contents for the TWTs, which means that the carriers in a given TWT must either all be in F_1 or all be in F_2, which can be expressed as follows:

$$\forall c, c' \in C, \ \mathbf{f}_c \in F \backslash F_2 \ \wedge \ \mathbf{f}_{c'} \in F \backslash F_1 \Rightarrow \mathbf{t}_c \neq \mathbf{t}_{c'} \qquad (11)$$

The objective is the minimization of the number of available TWTs actually used. That number n_{used} is a variable that can be obtained from the array $\mathbf{t}$ with two successive global counting constraints, the first one generating an array of the number of times each TWT is used, the second counting the number of non-zero values in the latter:

$$\min n_{\mathrm{used}} \qquad (12)$$

3.2 Multiprocessor Scheduling Part

The Scheduling Model. An analogy with multiprocessor scheduling problems is possible for the assignment of frequencies and polarizations, that is for the subproblem that only concerns the variable arrays $\mathbf{f}$, $\mathbf{p}$, and the constraints (2), (3), (4), (5) and (6). That problem, denoted by (S_1), is an extension of the model proposed in [6] where the frequency assignment is addressed regardless of the polarizations. Each beam $b \in B$ is assimilated to a single operation job whose processing time, expressed in time units, is non-preemptive and equal the number of carriers in that beam. Note that such a model is only valid because the frequencies of the carriers in a same beam are constrained by constraint (8) to be contiguous, the contiguousness of frequencies corresponding therefore to the non-preemptiveness of the processing times. Each maximal clique of G' is assimilated to a machine with non-overlapping constraints, while each maximal clique of G is associated to exactly two machines, one for each polarization. For each beam/job $b \in B$, $\mathcal{C}'_b$ denotes the set of machines that correspond to the cliques of G' that contain b, while $\mathcal{C}_{b,1}$ and $\mathcal{C}_{b,2}$ are the sets of machines representing the cliques of G containing b that are respectively associated to the polarizations 1 and 2. For constraint (4), it is assumed that the only restriction here is an upper-bound on the reuse factor $R \in \mathbb{N}^+$ of the channels (same bound for each channel), which leads to the definition of $M = \{m_1, \cdots, m_R\}$ identical parallel machines. Each job $b \in B$ requires simultaneously multiple machines. More precisely, it must be executed on:

- all the machines of $\mathcal{C}'_b$
- either all the machines of $\mathcal{C}_{b,1}$, or all the machines of $\mathcal{C}_{b,2}$
- one machine of M

Note that relying on cliques is not necessary to make this analogy with multi-processor scheduling, another option could be to define a machine for each binary constraint, but relying on cliques allows to take into account several constraints simultaneously, just like global constraints in constraint programming. In the example of Fig. 2, for the beam number 1 with the notations $\mathcal{C}'_1 = \{c'_{1,1}, c'_{1,2}\}$, $\mathcal{C}_{1,1} = \{c_{1,1,1}, c_{1,1,2}\}$ and $\mathcal{C}_{1,2} = \{c_{1,2,1}, c_{1,2,2}\}$, we have:

- $c'_{1,1}$ and $c'_{1,2}$ associated to the cliques/machines $\{1,2\}$ and $\{1,3\}$ of G'
- $c_{1,1,1}$ and $c_{1,1,2}$ associated to the machines of first polarization for the cliques $\{1,2,3\}$ and $\{1,3,4\}$ in G
- $c_{1,2,1}$ and $c_{1,2,2}$ associated to the machines of second polarization for the cliques $\{1,2,3\}$ and $\{1,3,4\}$ in G
- m_1 the machine in M used by the beam 1

In the example, the two carriers required in beam 1 use the second and third frequency channels and the first kind of polarization. With a common deadline for all the jobs being equal to the number of frequency channels N_F (equal to 4 in Fig. 2), one can see that solving this scheduling problem is equivalent to solving the considered subpart of our frequency assignment problem.

Proposition: (S_1) is equivalent to solving a multiprocessor scheduling problem, it is therefore NP-hard.

Proof: The parallel machine problem is a particular case of (S_1).   □

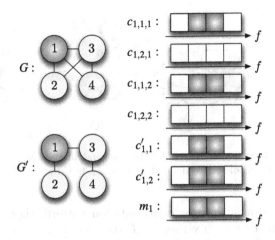

Fig. 2. Example of execution of one job on the machines.

Maximal Cliques Enumeration in Multibeam Satellites Interference Graphs. As explained in the previous paragraph, one promising direction to solve efficiently the scheduling part of the frequency assignment problem considered is to use the cliques of the interference graphs. It is thus of interest to study the theoretical and practical complexity of enumerating the maximal cliques. In multibeam systems, the analysis of their exhaustive enumeration differs depending on the type of graphs considered: regular layouts or random interference graphs.

Cliques In Regular Layouts. A regular layout is an organization of the beams that provides a continuous coverage of the zone with overlapping beams that describe an hexagonal lattice, as shown in Fig. 1 for instance. It is a very common choice for the system engineer since the contiguous coverage it provides can be a crucial specification of the customer, and also, it requires simpler antenna designs than a non-uniform layout. For a beam $b \in B$, let us denote by c_b the position of its center and by $\Gamma(b)$ the set of its adjacent beams. A common industrial approach for a regular layout with beams of radius r is to have $\Gamma(b) = \left\{ \tilde{b} \in B \mid \tilde{b} \neq b \text{ and } \|c_{\tilde{b}} - c_b\| < d \right\}$ with d being equal to either $3r$ or $2\sqrt{3}r$ leading to the representations (a) and (c) of Fig. 3. They are usually called 3-colors pattern and 4-colors pattern because with such edges in the interference graph, it is possible to partition the set of vertices into respectively 3 and 4 independent sets as shown in figure (b) and (d) of Fig. 3.

An important property of the regular interference graphs with the edges defined this way is the following:

Proposition: The maximal cliques of the interference graphs corresponding to the regular patterns in regular layouts can all be enumerated in polynomial time

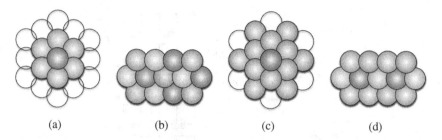

Fig. 3. (a) Adjacent beams, $3r$ threshold (b) Independent sets, $3r$ threshold (c) Adjacent beams, $2\sqrt{3}r$ threshold (d) Independent sets, $2\sqrt{3}r$ threshold.

Proof: The key idea is that for each exclusion pattern, there exists a finite number m such that for each vertex $b \in B$ there exist m potential cliques that might contain b, m being independent of the size N_B of the graph. For instance, for a graph with the edges of the 4-colors pattern, geometrical considerations allow to understand that, for a given vertex :

– it cannot belong to a clique of size 5 and more,
– the cliques of size 4 that might contain it are those of Fig. 4 plus those obtained by rotating of $\frac{\pi}{3}$ around the center of the corresponding beam leading to a total of 20 distinct potential cliques,
– the only way it can belong to a maximal clique of size 1, 2 or 3 is that the corresponding beam is surrounded by less beams than in the full configuration of Fig. 4, which can happen either because the beam in question is near the bound of the layout or because there are "holes" in it. Therefore, if such a clique exists, it is a subgraph of what would have been a clique of size 4 if some beams had not been missing. These situations are also in finite number and can be precisely enumerated.

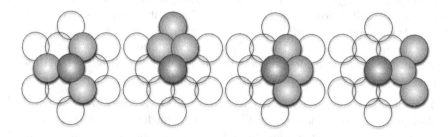

Fig. 4. Cliques of size 4 with 4-colors pattern.

Note that in the example of the 4-colors pattern, the number of cliques is therefore upper-bounded by $20N_B$. Each potential clique is characterized by a specific set of adjacent beams and, for the cliques of size less than 4, a set of non-existing beams whose positions are perfectly known in terms of distance to the beam tested and orientation with respect to a given reference direction, say the horizontal direction. The same type of rationale applies for the graphs defined with the 3-colors pattern. Therefore, to enumerate all the maximal cliques in the case of regular layouts, one would only have to iterate on the vertices $b \in B$, that is on the beams, and test each clique possibility to see which ones actually exist for each b. That way, the list of maximal cliques can gradually grow, simple tests allowing to avoid redundancies. In the end, the maximal cliques of the regular layouts are indeed enumerated with a polynomial complexity.   □

Cliques In Realistic Random Layouts. Even if the standard way to design a layout is to rely on the uniform patterns, it can be interesting to break that regularity in order to match the heterogeneity of the requirements over the service area. One can therefore have to work with a layout that can have beams of differents widths and positions for their centers that do not describe any particular known geometrical pattern. It was therefore necessary in that case to determine whether it was still an acceptable approach to enumerate the cliques before actually solving the frequency assignment problem. To do so, the slightly modified version of the Bron-Kerbosch [5] algorithm proposed by Tomita et al. [14] has been implemented and used on sets of graphs that were randomly generated with constraints on the vertex degrees. In practice, in multibeam satellite systems interference graphs, these vertex degrees are rarely less than 1 and greater than 12, so this has been specified as the main constraint in the constraint program used to generate these graphs. We generated 10000 different graphs of size $|B| = 200$ (maximum size for a realistic scenario) and observed that the mean number of cliques was 881 and the mean execution time was 14 milliseconds.

These cliques numbers are far from the $3^{\frac{|B|}{3}}$ upper bound of the number of cliques in an undirected graph, which is very interesting in practice because too high numbers of cliques could have made it impossible or unreasonable to rely on a model based on them. But most importantly, the computational times are relatively low, even instantaneous at the time scale of the designing phases of the satellite telecommunication systems. In the end, this means that this preliminary enumeration of the cliques is a pre-processing operation for the frequency assignment problem that is perfectly acceptable, whatever the type of layout.

3.3   Path Covering Part

Let us assume that the frequencies and the polarizations have been assigned somehow to the carriers of a given system, possibly with a scheduling based procedure as the one presented in Sect. 3.2. Then, one can wonder what the problem of assigning the TWTs to these carriers becomes, that problem being denoted by (S_2). The first important remark is that the constraint 11 on the type of TWTs can now be seen as additional incompatibilities in constraint 10 since the frequencies of the carriers are now known. The second is that it is now possible to represent the problem as a path-covering problem of a digraph in which the vertices represent the N_C carriers of the system (see Fig. 5), a path representing a TWT and its content. In this graph, for all $f \in F \backslash \{N_F\}$, the only possible direct successors of the carriers using the frequency f are those using the frequency $f + 1$, the in-degrees of the carriers using the frequency 1 being all equal to 0, just like the out-degrees of the carriers using the frequency N_F. As a consequence of these few properties, such graphs are acyclic. The incompatibilities between two carriers that cannot be in the same TWT/path are represented with dotted-line connections. For a given carrier, two situations impact the number of out-arcs: when this carrier is not the last carrier of the beam it belongs to, and when there exist incompatible carriers that use the next frequency. In the former case, only one arc leaves the carrier considered and its head is the next carrier in the corresponding beam. In the latter case, the carrier cannot be connected to the carriers with which an incompatibility is shared. Otherwise, for a carrier that is not in any of these two situations, it is connected to all the carriers using the next frequency. One can then see that assigning TWTs to the carriers comes down in that case to finding the minimum number of disjoint paths that cover all the vertices, the contiguity (constraint 8) and the fact that the same frequency cannot be used twice in a TWT (constraint 7) being automatically verified with a graph built that way. But there are also some additional constraints to take into account such as the upper-bound for the length of the paths (constraint 9), the constraint not to use the same TWT for two incompatible carriers (constraint 10), and finally the constraint that the carriers of a block of carriers must use the same TWT (constraint 1). In the end, an instance of the problem considered is entirely defined by: an acyclic digraph D whose vertices can be partitioned into a certain number of ordered "levels" and whose arcs are only between two vertices of a level and the next, an upper bound l for the length of the paths, a set for each carrier of the carriers it must share a TWT with (empty sets being allowed), and a set for each carrier of the carriers incompatible with that carrier (empty sets also allowed).

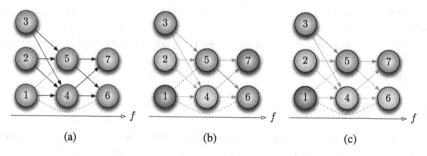

(a)   (b)   (c)

Fig. 5. (a) Instance of TWT-assignment graph (b) Suboptimal solution (c) Optimal solution.

Proposition: (S_2) is an NP-hard path-covering problem

Proof: Without the additional constraints (1,9,10), the problem of covering a digraph with a minimum number of point-disjoint paths can be solved in polynomial time as shown in [3]. But once they are taken into account, it can be proven that the problem becomes NP-hard. Indeed, let us consider an instance of the problem of finding a minimum cardinality cover of the elements of a partially ordered set (poset) with chains of restricted length, whose NP-completeness has been proven in [13]. It is common to represent that poset with a digraph partitioned in ordered levels, the edges connecting the comparable elements of the set from one level to the next: this is precisely a Hasse diagram. Then, with the upper bound for the path lengths equal to the maximum length of a chain and with, for each carrier, empty sets for the sets of carriers that must use the same TWT and the sets of incompatible carriers, one can see that solving this poset cover instance is equivalent to solving a particular instance of the path-covering problem considered in this paper. Therefore, it is also NP-complete. ☐

To solve it, the following integer linear programming model has been derived:

$$\min \sum_{t=1}^{N_T} u_t \tag{13}$$

s.t.

$$\forall c \in C, \sum_{t=1}^{N_T} x_{ct} = 1 \tag{14}$$

$$\forall t \in T, \forall f \in F, \sum_{c=1}^{N_C} y_{cf} x_{ct} \leq 1 \tag{15}$$

$$\forall t \in T, u_t \geq \frac{1}{N_C} \sum_{c=1}^{N_C} x_{ct} \tag{16}$$

$$\forall t \in T, \sum_{c=1}^{N_C} x_{ct} \leq n \tag{17}$$

$$\forall c, c' \in C \text{ that are incompatible}, \forall t \in T,$$
$$x_{ct} + x_{c't} \leq 1 \tag{18}$$

$$\forall c, c' \in C \text{ in the same block of carriers}, \forall t \in T,$$
$$x_{ct} - x_{c't} = 0 \tag{19}$$

$$\forall t \in T, \forall f \in F \backslash \{N_F\},$$
$$\sum_{c=1}^{N_C} \left(\left[\sum_{f'=f+1}^{N_F} y_{cf'} \right] + N_F y_{cf} - N_F y_{c(f+1)} \right) x_{ct} \leq N_F \tag{20}$$

where $y_{cf} \in \{0, 1\}$ are input Boolean arguments that indicate whether the carrier $c \in C$ uses the frequency $f \in F$, $x_{ct} \{0, 1\}$ are the Boolean variables that indicate if the carrier $c \in C$ uses the TWT $t \in T$, and finally the $u_t \in \{0, 1\}$ are the Boolean variables that indicate whether the TWT t is actually used. Constraint 14 is the constraint to have only one TWT assigned to each carrier, 15 forbids a given TWT to be used by two different carriers using the same frequency channel, 16 is the constraint that forces the u_t to be equal to 1 as soon as the TWT t is used at least once, 17 is the limit on the number of carriers in the same TWT, constraint 18 forbids two incompatible carriers to use the same TWT, 19 forces the carriers in the same block of carriers to use the same TWT, 20 ensures the contiguity of the frequency channels in each TWT, finally 13 is the minimization of the number of TWT actually used.

As mentioned in the constraints section, the contiguity of the frequencies in the traveling-wave tubes is a constraint that is not systematically required. Though, this constraint is the one that led to the notion of "path" in the carrier based directed graph. In the case where the contiguity is not required in the high power amplifiers, the problem turns out to be in fact a generalization of the multidimensional bin-packing problem. Indeed, the traveling-wave tubes can be seen as bins divided in N_F compartments, each compartment with capacity 1. Each carrier is a N_F-part object, each part having a size of 0 except for the part corresponding to the actual frequency of the carrier which has a size of 1. With additional constraints between carriers/objects that cannot be in the same bin, the minimization of the number of traveling-wave tubes comes down to packing the carriers/objects into as few TWT/bins as possible with each part going into the right compartment in a bin. In the end, the problem of allocating high power amplifier stays NP-hard. Note that in the experimental section that follows, the choice we made was to consider this contiguity constraint and, therefore, the associated path-covering problem.

4   Experimental Results

Experiments were needed to assess the performances of the two following approaches:

– Global Approach(GA):
   The global constraint program of Sect. 3.1 solved with a CP solver (Gecode)
– Decomposition Method(DM):
   Sequential solving of (S_1) of Sect. 3.2 with a CP solver (Gecode) and then of (S_2) of Sect. 3.3 with an ILP solver (Gurobi)

A first detailed example is presented in Fig. 6 with a fictitious scenario over France and Italy, with $N_B = 12$ regularly organized beams. The characteristics of the problem solved were the following:

– Each beam $b \in \{1, \cdots, 12\}$ of Fig. 6 has a required number of carriers n_b than is either equal to 1 or to 2, the carriers being indexed as shown inside the beams in Fig. 6a
– For the beams b with a number of carriers $n_b > 1$, we require contiguous carrier frequencies, same polarization and same TWT

- The system bandwidth is divided into $N_F = 6$ channels
- The acceptable frequency ranges for the TWTs are $\{1, 2, 3\}$ and $\{4, 5, 6\}$
- The TWT reuse upper-bound is set to 3, i.e. the width of an admissible frequency range
- The 4-color pattern is used to define binary interference constraints for the reuse of the same frequency-polarization couple (Fig. 3c)
- The 3-color pattern is used to define binary interference constraints for the reuse of the same frequency, regardless of the polarization (Fig. 3a)
- Carrier n°5 is incompatible with carriers n°9 and n°10, carrier n°13 is incompatible with carriers n°17 and n°18, carrier n°7 is incompatible with n°8, which means that they cannot use the same TWT
- Each frequency channel must be used at most third times
- Objective function: number of TWTs used

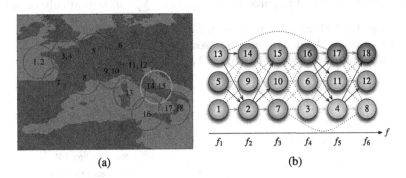

(a)   (b)

Fig. 6. (a) Multibeam coverage and polarizations (b) Frequencies and TWTs.

This is one of the instances for which GA solved with Gecode is unacceptably long to find a solution. On the other hand, with DM, the scheduling part and the subsequent binary linear program are both solved extremely efficiently respectively by Gecode and Gurobi. On Fig. 6a, the regular layout is reprensented with a ring color for each polarization, and on Fig. 6b, the frequencies of the carriers found in the scheduling part can be read on the horizontal axis, and each color for the carriers represents one TWT. Note that the design of Fig. 6 obtained for that example is optimal since the number of TWTs used is exactly equal to the number of carriers divided by the maximum number of carriers in a TWT.

When instances are randomly generated, note that there is no guarantee that they will be feasible. Even if this is true for both approaches, in the case of DM, this risk of infeasibility is even increased since some of the path-covering problem constraints are currently not anticipated in the preceding scheduling problem (the frequency ranges of the TWTs for instance). In practice, infeasibility is significantly harder to detect than actual solutions for feasible instances, at least when Gecode is used, that is in GA and in (S_1) of DM. In the results of this section, the statistic values presented only consider the instances that turned out to be feasible.

For each instance tested with the DM approach, the corresponding (S_1) scheduling problem is solved with Gecode using the corresponding subset of constraints in the global model of Sect. 3.1. Then, the solutions of (S_1) are transformed into (S_2) path-covering instances that are solved with Gurobi thanks to the ILP model we derived in Sect. 3.3. With GA, let us remind that the problem is entirely solved with Gecode. For the first phase of our series of experiments, we generated FAP instances with similar characteristics as the example detailed before, with the following few changes:

- Each beam $b \in \{1, \cdots, 12\}$ of Fig. 6 has a now required number of carriers n_b than is either equal to 0 or to 1
- The TWT carrier incompatibilities are now randomly generated (about 10 % of all the possible carrier couples)
- The overall number of required carriers $N_C = \sum_{1 \leq b \leq 12} n_b$ is gradually increased, from 4 to 12, 100 feasible instances being generated at each stage
- Each frequency channel cannot be used more than once when $4 \leq N_C \leq 6$ and more than twice when $7 \leq N_C \leq 12$

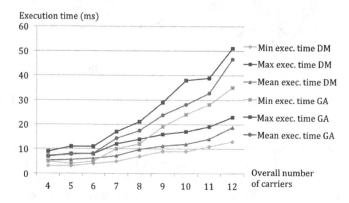

Fig. 7. Comparison of GA and DM on 4-carriers to 12-carriers instances with execution time statistics.

Table 1. Percentage of times the theoretical optimum is reached with DM for each set of instances of varying number of carriers in the system.

Number of carriers	4	5	6	7	8	9	10	11	12
Optimality of DM	87%	72%	75%	83%	59%	53%	69%	76%	64%

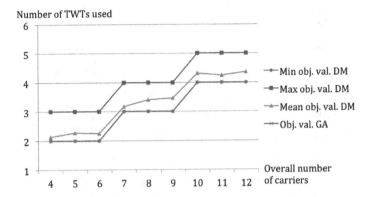

Fig. 8. Comparison of GA and DM on 4-carriers to 12-carriers instances with objective value statistics.

Figures 7 and 8 allow to compare GA and DM in terms of objective values and execution times. As expected, we can observe that in the case of a joint assignment of TWT, frequency and polarization to the carriers (GA), the execution times are greater than those of DM but the objective values are better in average. In the particular case of the instances we generated, GA always reaches the theoretical optimal value which is equal to

$$\text{ceiling} \left(\frac{\text{Overall number of carriers}}{\text{Maximum number of carriers in a TWT}} \right)$$

However, the decomposition method often manages to reach that optimal number of TWTs too as shown in Table 1. This is a crucial remark we wanted to emphasize since it is what legitimates the use of DM when GA is not usable in practice.

In the next phase of our experiments, the overall number of carriers in the system has been set to be greater than 12 and less than 19, the carrier requirements in each beam being either equal to 1 or 2, and the frequency channel reuse limit being now set to 3. As a result, some new constraints have to be taken into account for the beams b such that $n_b > 1$: contiguity of frequencies, same polarization and same TWT for the carriers belonging to the same beam. In practice, this is the point where GA becomes unusable both for feasible and infeasible instances, because of extremely long execution times even on these instances that are still relatively small compared to the biggest realistic situations. This explains why it has been necessary to develop DM. In Fig. 9, the execution times of (S_1) (scheduling) and (S_2) (path-covering) are compared on the whole range of instances, from 4-carriers instances to 18-carriers instances.

Two main things can be observed in that figure. First, the difference between the instances with at most 12 carriers and those with at least 13 carriers is clear: the new constraints linked to the beams for which the carrier requirement is strictly higher than one slow the search. Also, we see that the computational times grow faster for the scheduling problem than for the path-covering problem.

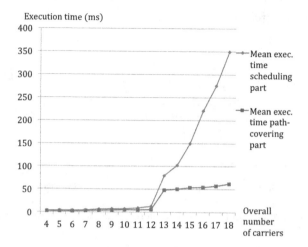

Fig. 9. (S_1) (scheduling part) and (S_2) (path-covering part) execution times.

That remark is even more important when we consider the fact that infeasible instances are also really hard to detect for Gecode in the scheduling part. (S_1) is therefore the subproblem that deserves more attention for future work, the goal being to solve the highest realistic instances. Our not yet exploited analysis of the cliques in the interference graphs could certainly be an interesting direction.

5   Conclusion

The models we proposed for this particular frequency assignment problem applied to the design of multibeam satellite systems allowed to algorithmically solve instances that could not be solved by satellite telecommunications engineers. We showed that the decomposition method we devised could produce solutions and even optimal solutions in reasonable computational times especially compared to the performances of the global constraint program for that problem. We also showed that relying on the cliques of the interference graphs was an acceptable direction and most likely a way to improve our current algorithms for the scheduling subproblem of our decomposition method. Concerning the path-covering problem, a series of experiments showed that realistic instances where solved almost instantaneously by the solver Gurobi, which tells us that we extracted an interesting subproblem, and we will definitely try to take advantage of this in some way in the next algorithms we will implement. To solve the largest realistic instances, work still has to be done to get faster results and improving the algorithms for the scheduling part might not be enough. Instead of solving the two identified subproblems sequentially, we might aim at more integrated approaches inspired by combinatorial Benders' cuts for instance, or with filtering algorithms solving locally the path covering problem.

References

1. Aardal, K.I., van Hoesel, S.P.M., Koster, A.M.C.A., Mannino, C., Sassano, A.: Models and solution techniques for frequency assignment problems. Ann. Oper. Res. **153**, 79–129 (2007)
2. Beldiceanu, N., Carlsson, M., Rampon, J.-X.: Global constraint catalog. SICS Research Report, May 2005
3. Boesch, F.T., Gimpel, J.F.: Covering the points of a digraph with point-disjoint paths and its application to code optimization. J. ACM **24**, 192–198 (1977)
4. Bousquet, M., Maral, G.: Satellite Communications Systems: Systems, Techniques and Technology, 5th edn, December 2009
5. Bron, C., Kerbosch, J.: Finding all cliques of an undirected graph. Commun. ACM **16**, 575–577 (1973)
6. Kiatmanaroj, K., Artigues, C., Houssin, L.: On scheduling models for the frequency interval assignment problem with cumulative interferences, August 2013
7. Koster, A., Tieves, M.: Column generation for frequency assignment in slow frequency hopping networks. EURASIP J. Wirel. Commun. Networking **2012**(1), 1–14 (2012)
8. Luna, F., Estbanez, C., Len, C., Chaves-Gonzlez, J., Nebro, A., Aler, R., Gmez-Pulido, J.: Optimization algorithms for large-scale real-world instances of the frequency assignment problem. Soft Comput. **15**(5), 975–990 (2011)
9. Muoz, D.: Algorithms for the generalized weighted frequency assignment problem. Comput. Oper. Res. **39**(12), 3256–3266 (2012)
10. Salma, A., Ahmad, I., Omran, M., Mohammad, M.: Frequency assignment problem in satellite communications using differential evolution. Comput. Oper. Res. **37**(12), 2152–2163 (2010)
11. Schulte, C., Tack, G., Lagerkvist, M.Z.: Modeling and programming with gecode, 522 (2013)
12. Segura, C., Miranda, G., Len, C.: Parallel hyperheuristics for the frequency assignment problem. Memetic Comput. **3**(1), 33–49 (2011)
13. Shum, H., Trotter, L.E.: Cardinality-restricted chains and antichains in partially ordered sets. Discrete Appl. Math. **65**, 421–439 (1996)
14. Tomita, E., Tanaka, A., Takahashi, H.: The worst-case time complexity for generating all maximal cliques and computational experiments. Theor. Comput. Sci. **363**, 28–42 (2006)
15. van Hoeve, W.-J.: The all different constraint: a survey, Cornell University Library (2001)
16. Wang, J., Cai, Y.: Multiobjective evolutionary algorithm for frequency assignment problem in satellite communications. Soft Comput. **19**(5), 1–25 (2014)
17. Yang, C., Peng, S., Jiang, B., Wang, L., Li, R.: Hyper-heuristic genetic algorithm for solving frequency assignment problem in TD-SCDMA. In: Proceedings of the 2014 Conference Companion on Genetic and Evolutionary Computation Companion, pp. 1231–1238. ACM (2014)

Selection-Based Approach to Cooperative Interval Games

Jan Bok[1]([✉]) and Milan Hladík[2]

[1] Computer Science Institute, Charles University in Prague,
Malostranské náměstí 25, 11800 Prague, Czech Republic
bok@iuuk.mff.cuni.cz
[2] Department of Applied Mathematics, Faculty of Mathematics and Physics,
Charles University in Prague, Malostranské náměstí 25,
11800 Prague, Czech Republic
hladik@kam.mff.cuni.cz

Abstract. Cooperative interval games are a generalized model of cooperative games in which worth of every coalition corresponds to a closed interval representing the possible outcomes of its cooperation. Selections are all possible outcomes of the interval game with no additional uncertainty. We introduce new selection-based classes of interval games and prove their characterization theorems and relations to existing classes based on the interval weakly better operator. We show a new results regarding the core and imputations and examine a problem of equality of two different versions of core, which is the main stability solution of cooperative games. Then we introduce definition of strong imputation and strong core as a universal solution concept of interval games.

1 Introduction

Uncertainty and inaccurate data are issues occurring very often in the real world situations. Therefore it is important to be able to make decisions even when the exact data are not available and only bounds on them are known.

In classical cooperative game theory, every group of players (*coalition*) knows precise reward for their cooperation; in cooperative interval games, only the worst and the best possible outcome are known. Such situations can be naturally modeled with intervals encapsulating these outcomes.

Cooperation under interval uncertainty was first considered by Branzei, Dimitrov and Tijs in 2003 to study bankruptcy situations [1] and later further extensively studied by Alparslan Gök in her PhD thesis [2] and other papers written together with Branzei et al. (see the references section of [3] for more).

However, their approach is almost exclusively aimed on interval solutions, that is on payoff distributions consisting of intervals and thus containing another uncertainty. This is in contrast with selections – possible outcomes of an interval game with no additional uncertainty. Selection-based approach was never systematically studied and not very much is known. This paper is trying to fix

© Springer International Publishing Switzerland 2015
D. de Werra et al. (Eds.): ICORES 2015, CCIS 577, pp. 40–53, 2015.
DOI: 10.1007/978-3-319-27680-9_3

this and summarizes our results regarding a selection-based approach to interval games.

This paper has the following structure. Section 2 is a preliminary section which concisely presents necessary definitions and facts on classical cooperative games, interval analysis and cooperative interval games. Section 3 is devoted to new selection-based classes of interval games. We consequently prove their characterizations and relations to existing classes. Section 4 focuses on the so called core incidence problem which asks under which conditions are the selection core and the set of payoffs generated by the interval core equal. In Sect. 5, definitions of strong core and strong imputation are introduced as new concepts. We show some remarks on strong core, one of them being a characterization of games with the strong imputation and strong core. Finally, we conclude this paper with the summary of our results and possible directions for future research.

On Mathematical Notation

- We will use $\leq$ relation on real vectors. For every $x, y \in \mathbb{R}^N$ we write $x \leq y$ if $x_i \leq y_i$ holds for every $1 \leq i \leq N$.
- We do not use symbol $\subset$ in this paper. Instead, $\subseteq$ and $\subsetneq$ are used for subset and proper subset, respectively, to avoid ambiguity.

2   Preliminaries

2.1   Classical Cooperative Game Theory

Comprehensive sources on classical cooperative game theory are for example [4–7]. For more info about on applications, see e.g. [8–10]. Here we present only necessary background theory for study of interval games. We examine games with transferable utility (TU), therefore by cooperative game we mean cooperative TU game.

Definition 1 (Cooperative Game). *Cooperative game is an ordered pair* (N, v), *where* $N = \{1, 2, \ldots, n\}$ *is a set of players and* $v : 2^N \to \mathbb{R}$ *is a characteristic function of the cooperative game. We further assume that* $v(\emptyset) = 0$.

The set of all cooperative games with player set N is denoted by G^N.

Subsets of N are called *coalitions* and N itself is called a *grand coalition*.

We often write v instead of (N, v), because we can easily identify game with its characteristic function without loss of generality.

To further analyze players' gains, we will need a *payoff vector* which can be interpreted as a proposed distribution of reward between players.

Definition 2 (Payoff Vector). *Payoff vector for a cooperative game* (N, v) *is a vector* $x \in \mathbb{R}^N$ *with* x_i *denoting reward given to ith player.*

Definition 3 (Imputation). *An imputation of* $(N, v) \in G^N$ *is a vector* $x \in \mathbb{R}^N$ *such that* $\sum_{i \in N} x_i = v(N)$ *and* $x_i \geq v(\{i\})$ *for every* $i \in N$.

The set of all imputations of a given cooperative game (N, v) *is denoted by* $I(v)$.

Definition 4 (Core). *The core of* $(N, v) \in G^N$ *is the set*

$$C(v) = \left\{ x \in I(v), \sum_{i \in S} x_i \geq v(S), \forall S \subseteq N \right\}.$$

There are many important classes of cooperative games. Here we show the most important ones.

Definition 5 (Monotonic Game). *A game* (N, v) *is monotonic if for every* $T \subseteq S \subseteq N$ *we have*

$$v(T) \leq v(S).$$

Informally, in monotonic games, bigger coalitions are stronger.

Definition 6 (Superadditive Game). *A game* (N, v) *is superadditive if for every* $S, T \subseteq N$, $S \cap T = \emptyset$ *we have*

$$v(T) + v(S) \leq v(S \cup T).$$

In a superadditive game, coalition has no incentive to divide itself, since together they will always achieve at least as much as separated.

Superadditive game is not necessarily monotonic. Conversely, monotonic game is not necessarily superadditive. However, these classes have a nonempty intersection. Check Caulier's paper [11] for more details on relation of these two classes.

Definition 7 (Additive Game). *A game* (N, v) *is additive if for every* $S, T \subseteq N$, $S \cap T = \emptyset$ *we have*

$$v(T) + v(S) = v(S \cup T).$$

Observe that additive games are superadditive as well.

Another important type of game is a *convex game*.

Definition 8 (Convex Game). *A game* (N, v) *is convex if its characteristic function is supermodular. The characteristic function is supermodular if for every* $S \subseteq T \subseteq N$ *holds*

$$v(T) + v(S) \leq v(S \cup T) + v(S \cap T).$$

Clearly, supermodularity implies superadditivity.

Convex games have many nice properties. we show the most important one.

Theorem 1. *(Shapley 1971 [12]) If a game* (N, v) *is convex, then its core is nonempty.*

2.2   Interval Analysis

Definition 9 (Interval). *The interval X is a set*

$$X := [\underline{X}, \overline{X}] = \{x \in \mathbb{R} : \underline{X} \le x \le \overline{X}\}.$$

With $\underline{X}$ being the lower bound and $\overline{X}$ being the upper bound of the interval.

From now on, when we say an interval we mean a closed interval. The set of all real intervals is denoted by $\mathbb{IR}$.

The following definition shows how to do a basic arithmetics with intervals [13].

Definition 10. *For every $X, Y, Z \in \mathbb{IR}$ and $0 \notin Z$ define*

$$X + Y := [\underline{X} + \underline{Y}, \overline{X} + \overline{Y}],$$
$$X - Y := [\underline{X} - \overline{Y}, \overline{X} - \underline{Y}],$$
$$X \cdot Y := [\min S, \max S], \; S = \{\underline{X}\,\overline{Y}, \overline{X}\,\underline{Y}, \underline{X}\,\underline{Y}, \overline{X}\,\overline{Y}\},$$
$$X / Z := [\min S, \max S], \; S = \{\underline{X}/\overline{Z}, \overline{X}/\underline{Z}, \underline{X}/\underline{Z}, \overline{X}/\overline{Z}\}.$$

2.3   Cooperative Interval Games

Definition 11. (Cooperative Interval Game). *A cooperative game is an ordered pair (N, w), where $N = \{1, 2, \dots, n\}$ is a set of players and $w : 2^N \to \mathbb{IR}$ is a characteristic function of the cooperative game. We further assume that $w(\emptyset) = [0, 0]$.*

The set of all interval cooperative games on player set N is denoted by IG^N.

We often write $w(i)$ instead of $w(\{i\})$.

Remark 1. Each cooperative interval game in which the characteristic function maps to degenerate intervals only can be associated with some classical cooperative game. Converse holds as well.

Definition 12 (Border Games). *For every $(N, w) \in IG^N$, border games $(N, \underline{w}) \in G^N$ (lower border game) and $(N, \overline{w}) \in G^N$ (upper border game) are given by $\underline{w}(S) = \underline{w(S)}$ and $\overline{w}(S) = \overline{w(S)}$ for every $S \in 2^N$.*

Definition 13 (Length Game). *The length game of $(N, w) \in IG^N$ is the game $(N, |w|) \in G^N$ with*

$$|w|(S) = \overline{w}(S) - \underline{w}(S), \; \forall S \in 2^N.$$

The basic notion of our approach will be a selection and consequently a selection imputation and a selection core.

Definition 14 (Selection). *A game $(N, v) \in G^N$ is a selection of $(N, w) \in IG^N$ if for every $S \in 2^N$ we have $v(S) \in w(S)$. Set of all selections of (N, w) is denoted by $\mathrm{Sel}(w)$.*

Note that border games are particular examples of selections.

Definition 15 (Interval Selection Imputation). *The set of interval selection imputations (or just selection imputations) of* $(N, w) \in IG^N$ *is defined as*

$$\mathcal{SI}(w) = \bigcup \{ I(v) \mid v \in \text{Sel}(w) \}.$$

Definition 16 (Interval Selection Core). *Interval selection core (or just selection core) of* $(N, w) \in IG^N$ *is defined as*

$$\mathcal{SC}(w) = \bigcup \{ C(v) \mid v \in \text{Sel}(w) \}.$$

Alparslan Gök [2] choose an approach using a weakly better operator. That was inspired by [14].

Definition 17 (Weakly better Operator $\succeq$). *Interval I is weakly better than interval J ($I \succeq J$) if $\underline{I} \geq \underline{J}$ and $\overline{I} \geq \overline{J}$. Furthermore, $I \preceq J$ if and only if $\underline{I} \leq \underline{J}$ and $\overline{I} \leq \overline{J}$. Interval I is better than J ($I \succ J$) if and only if $I \succeq J$ and $I \neq J$.*

Their definition of imputation and core is as follows.

Definition 18 (Interval Imputation). *The set of interval imputations of* $(N, w) \in IG^N$ *is defined as*

$$\mathcal{I}(w) := \left\{ (I_1, I_2, \dots, I_N) \in \mathbb{IR}^N \mid \sum_{i \in N} I_i = w(N),\ I_i \succeq w(i),\ \forall i \in N \right\}.$$

Definition 19 (Interval Core). *An interval core of* $(N, w) \in IG^N$ *is defined as*

$$\mathcal{C}(w) := \left\{ (I_1, I_2, \dots, I_N) \in \mathcal{I}(w) \mid \sum_{i \in S} I_i \succeq w(S),\ \forall S \in 2^N \setminus \{\emptyset\} \right\}.$$

Important difference between definitions of interval and selection core and imputation is that selection concepts yield a payoff vectors from $\mathbb{R}^N$, while $\mathcal{I}$ and $\mathcal{C}$ yield vectors from $\mathbb{IR}^N$.

(Notation) Throughout the papers on cooperative interval games, notation, especially of core and imputations, is not unified. It is therefore possible to encounter different notation from ours.

Also, in these papers, selection core is called core of interval game. We consider that confusing and that is why do we use term selection core instead. The term selection imputation is used because of its connection with selection core.

The following classes of interval games have been studied earlier (see e.g. [15]).

Definition 20 (Size Monotonicity). *A game $(N, w) \in IG^N$ is size monotonic if for every $T \subseteq S \subseteq N$ we have*

$$|w|(T) \leq |w|(S).$$

That is when its length game is monotonic.

The class of size monotonic games on player set N is denoted by SMIG^N.

As we can see, size monotonic games capture situations in which an interval uncertainty grows with the size of coalition.

Definition 21 (Superadditive Interval Game). *A game* $(N, w) \in IG^N$ *is superadditive interval game if for every* $S, T \subseteq N$, $S \cap T = \emptyset$ *holds*

$$w(T) + w(S) \preceq w(S \cup T),$$

and its length game is superadditive. We denote by SIG^N *class of superadditive interval games on player set* N.

We should be careful with the following analogy of convex game, since unlike for classical games, supermodularity is not the same as convexity.

Definition 22 (Supermodular Interval Game). *An interval game* (N, v) *is supermodular interval if for every* $S \subseteq T \subseteq N$ *holds*

$$v(T) + v(S) \preceq v(S \cup T) + v(S \cap T).$$

We get immediately that interval game is supermodular interval if and only if its border games are convex.

Definition 23 (Convex Interval Game). *An interval game* (N, v) *is convex interval if its border games and length game are convex.*
We write CIG^N *for a set of convex interval games on player set* N.

Convex interval game is supermodular as well but converse does not hold in general. See [15] for characterizations of convex interval games and discussion on their properties.

3 Selection-Based Classes of Interval Games

We will now introduce new classes of interval games based on the properties of their selections. We think that it is natural way to generalize special classes from classical cooperative game theory. Consequently, we show their characterizations and relation to classes from preceding subsection.

Definition 24 (Selection Monotonic Interval Game). *An interval game* (N, v) *is selection monotonic if all its selections are monotonic games. The class of such games on player set* N *is denoted by* SeMIG^N.

Definition 25 (Selection Superadditive Interval Game). *An interval game* (N, v) *is selection superadditive if all its selections are superadditive games. The class of such games on player set* N *is denoted by* SeSIG^N.

Definition 26 (Selection Convex Interval Game). *An interval game* (N, v) *is selection convex if all its selections are convex games. The class of such games on player set* N *is denoted by* SeCIG^N.

We see that many properties persist. For example, a selection convex game is a selection superadditive as well. Selection monotonic and selection superadditive are not subset of each other but their intersection is nonempty. Furthermore, selection core of selection convex game is nonempty, which is an easy observation.

We will now show characterizations of these three classes and consequently show their relations to existing classes presented in Subsect. 2.3.

Theorem 2. *An interval game* (N, w) *is selection monotonic if and only if for every* $S, T \in 2^N$, $S \subsetneq T$ *holds*

$$\overline{w}(S) \leq \underline{w}(T).$$

Proof. For the "only if" part, suppose that (N, w) is a selection monotonic and $\overline{w}(S) > \underline{w}(T)$ for some $S, T \in 2^N$, $S \subsetneq T$. Then selection (N, v) with $v(S) = \overline{w}(S)$ and $v(T) = \underline{w}(T)$ clearly violates monotonicity and we arrive at a contradiction with assumptions.

Now for the "if" part. For any two subsets S, T of N, one of the situations $S \subsetneq T$, $T \subsetneq S$ or $S = T$ occur. For $S = T$, in every selection v, $v(S) \leq v(S)$ holds. As for the other two situations, it is obvious that monotonicity cannot be violated as well since $v(S) \leq \overline{w}(S) \leq \underline{w}(T) \leq v(T)$. □

Notice the importance of using $S \subsetneq T$ in the formulation of Theorem 2. That is because using of $S \subseteq T$ (thus allowing situation $S = T$) would imply $\overline{w}(S) \leq \underline{w}(S)$ for every S in selection monotonic game which is obviously not true in general. In characterizations of selection superadditive and selection convex games, similar situation arises.

Theorem 3. *An interval game* (N, w) *is selection superadditive if and only if for every* $S, T \in 2^N$ *such that* $S \cap T = \emptyset$, $S \neq \emptyset$, $T \neq \emptyset$ *holds*

$$\overline{w}(S) + \overline{w}(T) \leq \underline{w}(S \cup T).$$

Proof. Similar to proof of Theorem 2. □

We give a characterization of selection convex games as well

Theorem 4. *An interval game* (N, w) *is selection convex if and only if for every* $S, T \in 2^N$ *such that* $S \not\subseteq T$, $T \not\subseteq S$, $S \neq \emptyset$, $T \neq \emptyset$ *holds*

$$\overline{w}(S) + \overline{w}(T) \leq \underline{w}(S \cup T) + \underline{w}(S \cap T).$$

Proof. Similar to proof of Proposition 2.

Now let us look on a relation with existing classes of interval games.

For selection monotonic and size monotonic games, their relation is obvious. For nontrivial games (that is games with the player set size greater than one), a selection monotonic game is not necessarily size monotonic. Converse is the same.

Theorem 5. *For every player set N with $|N| > 1$, the following assertions hold.*

(i) $\mathrm{SeSIG}^N \not\subseteq \mathrm{SIG}^N$.
(ii) $\mathrm{SIG}^N \not\subseteq \mathrm{SeSIG}^N$.
(iii) $\mathrm{SeSIG}^N \cap \mathrm{SIG}^N \neq \emptyset$.

Proof. In *(i)*, we can construct the counterexample in the following way.

Let us construct game (N, w). For $w(\emptyset)$, interval is given. Now for any non-empty coalition, set $w(S) := [2|S| - 2, 2|S| - 1]$. For any $S, T \in 2^N$ with S and T being nonempty, the following holds with the fact that $|S| + |T| = |S \cup T|$ taken into account.

$$\overline{w}(S) + \overline{w}(T) = (2|S| - 1) + (2|T| - 1)$$
$$= 2|S \cup T| - 2$$
$$= \underline{w}(S \cup T)$$

So (N, w) is selection superadditive by Theorem 3. Its length game, however, is not superadditive, since for any two nonempty coalitions with the empty intersection $|w|(S) + |w|(T) = 2 \not\leq 1 = |w|(S \cup T)$ holds.

In *(ii)*, we can construct the following counterexample (N, w'). Set $w'(S) = [0, |S|]$ for any nonempty S. The lower border game is trivially superadditive. For the upper game, $\overline{w'}(S) + \overline{w'}(T) = |S| + |T| = |S \cup T| = \overline{w'}(S \cup T)$ for any S, T with the empty intersection, so the upper game is superadditive. Observe that the length game is the same as the upper border game. This shows interval superadditivity.

However, (N, w') is clearly not selection superadditive because of nonzero upper bounds, zero lower bounds of nonempty coalitions and the characterization of SeSIG^N taken into account.

(iii) Nonempty intersection can be argued easily by taking some super-additive game $(N, c) \in G^N$. Then we can define corresponding game $(N, d) \in IG^N$ with

$$d(S) = [c(S), c(S)], \quad \forall S \in 2^N.$$

Game (N, d) is selection superadditive since its only selection is (N, c). And it is superadditive interval game since border games are supermodular and length game is $|w|(S) = 0$ for every coalition, which trivially implies its super-additivity. □

Theorem 6. *For every player set N with $|N| > 1$, following assertions hold.*

(i) $\mathrm{SeCIG}^N \not\subseteq \mathrm{CIG}^N$.
(ii) $\mathrm{CIG}^N \not\subseteq \mathrm{SeCIG}^N$.
(iii) $\mathrm{SeCIG}^N \cap \mathrm{CIG}^N \neq \emptyset$.

Proof. For *(i)*, take a game (N, w) assigning to each nonempty coalition S interval $[2^{|S|} - 2, 2^{|S|} - 1]$. From Theorem 4, we get that for inequalities which must hold in order to meet necessary conditions of game to be selection convex, $|S| < |S \cup T|$ and $|T| < |S \cup T|$ must hold. That gives the following inequality.

$$\overline{w}(S) + \overline{w}(T) \leq (2^{|S \cup T|-1} - 1) + (2^{|S \cup T|-1} - 1)$$
$$= 2^{|S \cup T|} - 2$$
$$= \underline{w}(S \cup T)$$
$$\leq \underline{w}(S \cup T) + \underline{w}(S \cap T)$$

This concludes that (N, w) is selection convex. We see that the border games and the length game are convex too. To have a game so that it is selection convex and not convex interval game, we can take (N, c) and set $c(S) := w(S)$ for $S \neq N$ and $v(N) := [\underline{w}(S), \underline{w}(S)]$. Now the game (N, c) is still selection convex, but its length game is not convex, so (N, v) is not convex interval game, which is what we wanted.

In *(ii)*, we can take a game (N, w') from the proof of Theorem 5 *(ii)*. From the fact that $|S| + |T| = |S \cup T| + |S \cap T|$, it is clear that $\overline{w'}$ is convex. The lower border game is trivially convex and the length game is the same as upper. However, for nonempty $S, T \in 2^N$ such that $S \not\subseteq T$, $T \not\subseteq S$, $S \neq \emptyset$, $T \neq \emptyset$, convex selection games characterization is clearly violated.

As for *(iii)*, we can use the same steps as in *(iii)* of Theorem 5 or we can use a game (N, w) from *(i)* of this theorem.   □

4   Core Coincidence

In Alparslan-Gök's PhD thesis [2,16], the following question is suggested:

> *"A difficult topic might be to analyze under which conditions the set of payoff vectors generated by the interval core of a cooperative interval game coincides with the core of the game in terms of selections of the interval game."*

We decided to examine this topic. We call it a *core coincidence problem*. This section shows our results.

We remind the reader that whenever we talk about relation and maximum, minimum, maximal, minimal vectors, we mean relation $\leq$ on real vectors unless we say otherwise.

The main thing to notice is that while the interval core gives us a set of interval vectors, selection core gives us a set of real numbered vectors. To be able to compare them, we need to assign to a set of interval vectors a set of real vectors generated by these interval vectors. That is exactly what the following function gen does.

Definition 27. *The function* gen $: 2^{\mathbb{IR}^N} \to 2^{\mathbb{R}^N}$ *maps to every set of interval vectors a set of real vectors. It is defined as*

$$\text{gen}(S) = \bigcup_{s \in S} \left\{ (x_1, x_2, \ldots, x_n) \mid x_i \in s_i, \ s \in \mathbb{IR}^N \right\}.$$

Core coincidence problem can be formulated as this: What are the necessary and sufficient condition to satisfy $\text{gen}(\mathcal{C}(w)) = \mathcal{SC}(w)$?

The main results of this subsection are the two following theorems which give an answer to the aforementioned question.

In the following text, by mixed system we mean a system of equalities and inequalities.

Theorem 7. *For every interval game (M, w) we have $\text{gen}(\mathcal{C}(w)) \subseteq \mathcal{SC}(w)$.*

Proof. For any $x \in \text{gen}(\mathcal{C}(w))$, inequality $\underline{v}(N) \leq \sum_{i \in N} x_i \leq \overline{v}(N)$ obviously holds. Furthermore, x is in the core for any selection of the interval game (N, s) with s given by

$$s(S) = \begin{cases} \left[\sum_{i \in N} x_i, \sum_{i \in N} x_i\right] \text{ if } S = N, \\ \left[\underline{w}(S), \min(\sum_{i \in S} x_i, \overline{w}(S))\right] \text{ otherwise.} \end{cases}$$

Clearly, $\text{Sel}(s) \subseteq \text{Sel}(w)$ and $\text{Sel}(s) \neq \emptyset$. That concludes $\text{gen}(\mathcal{C}(w)) \subseteq \mathcal{SC}(w)$. □

Theorem 8. (Core coincidence characterization) *For every interval game (N, w) we have $\text{gen}(\mathcal{C}(w)) = \mathcal{SC}(w)$ if and only if for every $x \in \mathcal{SC}(w)$ there exist non-negative vectors $l^{(x)}$ and $u^{(x)}$ such that*

$$\sum_{i \in N}(x_i - l_i^{(x)}) = \underline{w}(N), \tag{4.1}$$

$$\sum_{i \in N}(x_i + u_i^{(x)}) = \overline{w}(N), \tag{4.2}$$

$$\sum_{i \in S}(x_i - l_i^{(x)}) \geq \underline{w}(S), \ \forall S \in 2^N \setminus \{\emptyset\}, \tag{4.3}$$

$$\sum_{i \in S})x_i + u_i^{(x)}) \geq \overline{w}(S), \ \forall S \in 2^N \setminus \{\emptyset\}. \tag{4.4}$$

Proof. First, we observe that with Theorem 7 taken into account, we only need to take care of $\text{gen}(\mathcal{C}(w)) \supseteq \mathcal{SC}(w)$ to obtain equality.

For $\text{gen}(\mathcal{C}(w)) \supseteq \mathcal{SC}(w)$, suppose we have some $x \in \mathcal{SC}(w)$. For this vector, we need to find some interval $X \in \mathcal{C}(w)$ such that $x \in \text{gen}(X)$. This is equivalent to the task of finding two nonnegative vectors $l^{(x)}$ and $u^{(x)}$ such that

$$([x_1 - l_1^{(x)}, x_1 + u_1^{(x)}]), [x_2 - l_2^{(x)}, x_2 + u_2^{(x)}], \ldots, [x_n - l_n^{(x)}, x_n + u_n^{(x)}]) \in \mathcal{C}(w).$$

From the definition of interval core, we can see that these two vectors have to satisfy exactly the mixed system (4.1)–(4.4). That completes the proof. □

Example 1. Consider an interval game with $N = \{1, 2\}$ and $w(\{1\}) = w(\{2\}) = [1, 3]$ and $w(N) = [1, 4]$. Then vector $(2, 2)$ lies in the core of the selection with $v(\{1\}) = v(\{2\}) = 2$ and $v(N) = 4$. However, to satisfy Eq. (4.1), we need to have $\sum_{i \in N} l_i = 3$ which means that either l_1 or l_2 has to be greater than 1. That means we cannot satisfy (4.3) and we conclude that $\text{gen}(\mathcal{C}(w)) \neq \mathcal{SC}(w)$.

The following theorem shows that it suffices to check only minimal and maximal vectors of $SC(w)$.

Theorem 9. *For every interval game (N, w), if there exist vectors $q, r, x \in \mathbb{R}^N$ such that $q, r \in \text{gen}(\mathcal{C}(w))$ and $q_i \leq x_i \leq r_i$ for every $i \in N$, then $x \in \text{gen}(\mathcal{C}(w))$.*

Proof. Let $l^{(r)}, u^{(r)}, l^{(q)}, u^{(q)}$ be the corresponding vectors in sense of Theorem 8. We need to find vectors $l^{(x)}$ and $u^{(x)}$ satisfying (4.1)–(4.4) of Theorem 8.
Let's define vectors $dq, dr \in \mathbb{R}^N$:

$$dq_i = x_i - q_i,$$

$$dr_i = r_i - x_i.$$

Finally, we define $l^{(x)}$ and $u^{(x)}$ in this way:

$$l_i^{(x)} = dq_i + l_i^{(q)},$$
$$u_i^{(x)} = dr_i + u_i^{(r)}.$$

We need to check that we satisfy (4.1)–(4.4) for x, $l^{(x)}$ and $u^{(x)}$ We will show (4.2) only, since remaining ones can be done in a similar way.

$$\sum_{i \in N}(x_i - l_i^{(x)}) = \sum_{i \in N}(x_i - dq_i - l_i^{(q)})$$
$$= \sum_{i \in N}(x_i - x_i + q_i - l_i^{(q)})$$
$$= \sum_{i \in N}(q_i - l_i^{(q)})$$
$$= \underline{w}(N).$$

$\square$

For games with additive border games (see Definition 7), we got a following result.

Theorem 10. *For an interval game (N, w) with additive border games, the payoff vector $(\underline{w}(1), \underline{w}(2), \ldots, \underline{w}(n)) \in \text{gen}(\mathcal{C}(w))$.*

Proof. First, let us look on an arbitrary additive game (A, v_A). From additivity condition and the fact that we can write any subset of A as a union of one-player sets we conclude that $v_A(A) = \bigcup_{i \in A} v_A(\{i\})$ for every coalition A. This implies that vector a with $a_i = v_A(\{i\})$ is in the core.
This argument can be applied to border games of (N, w). The vector $q \in \mathbb{R}^N$ with $q_i = \underline{w}(i)$ is an element of the core of $(N, \underline{w})$ and an element of $SC(w)$.
For the vector q we want to satisfy the mixed system (4.1)–(4.4) of Theorem 8.
Take the vector l containing zeros only and the vector u with $u_i = |w|(i)$. From the additivity, we get that $\sum_{i \in N} q_i - l_i = \underline{w}(N)$ and $\sum_{i \in N} q_i + u_i = \overline{w}(N)$.
Additivity further implies that inequalities (4.3) and (4.4) hold for q, l and u. Therefore, q is an element of $\text{gen}(\mathcal{C}(w))$.

$\square$

Theorem 10 implies that for games with additive border games, we need to check the existence of vectors l and u from (4.1)–(4.4) of Theorem 8 for maximal vectors of $\mathcal{SC}$ only. That follows from the fact that for any vector $y \in \mathcal{SC}(w)$ holds $(\underline{w}(1), \underline{w}(2), \ldots, \underline{w}(n)) \leq y$. In other words, $(\underline{w}(1), \underline{w}(2), \ldots, \underline{w}(n))$ is a minimum vector of $\mathcal{SC}(w)$.

5   Strong Imputation and Core

In this subsection, our focus will be on a new concept of *strong imputation* and *strong core*.

Definition 28 (Strong Imputation). *For a game $(N, w) \in IG^N$ a strong imputation is a vector $x \in \mathbb{R}^N$ such that x is an imputation for every selection of (N, w).*

Definition 29 (Strong Core). *For a game $(N, w) \in IG^N$ the strong core is the union of vectors $x \in \mathbb{R}^N$ such that x is an element of core of every selection of (N, w).*

Strong imputation and strong core can be considered as somewhat "universal" solutions. We show the following three simple facts about the strong core.

Theorem 11. *For every interval game with nonempty strong core, $w(N)$ is a degenerate interval.*

Proof. Easily by the fact that an element c of strong core must be efficient for every selection and therefore $\sum_{i \in N} c_i = \underline{w}(N) = \overline{w}(N)$.   □

This leads us to characterizing games with nonempty strong core.

Theorem 12. *An interval game (N, w) has a nonempty strong core if and only if $w(N)$ is a degenerate interval and the upper game $\overline{w}$ has a nonempty core.*

Proof. Combination of Theorem 11 and the fact that an element c of the strong core has to satisfy $\sum_{i \in S} c_i \geq v(S)$, $\forall v \in \mathrm{Sel}(w)$, $\forall S \in 2^N \setminus \emptyset$. We see that this fact is equivalent to condition $\sum_{i \in S} c_i \geq \overline{w}(S)$, $\forall S \in 2^N \setminus \emptyset$. Proving an equivalence is then straightforward.   □

We observe that we can easily derive a characterization of games with nonempty strong imputation set.

Strong core also has the following important property.

Theorem 13. *For every element c of the strong core of (N, w), $c \in \mathrm{gen}(\mathcal{C}(w))$.*

Proof. The vector c has to satisfy mixed system (4.1)–(4.4) of Theorem 8 for some $l, u \in \mathbb{IR}^N$. We show that $l_i = u_i = 0$ will do the thing.

Equations (4.1) and (4.2) are satisfied by taking Theorem 11 into account. Inequalities (4.3) and (4.4) are satisfied as the consequence of Theorem 12.   □

The reason behind the using of name strong core and strong imputation comes form the interval linear algebra, where strong solutions of interval system are a solutions for any realization (selection) of interval matrices A and b in $Ax = b$.

One could say why we did not introduce a strong game as game in which each of its selection does have an nonempty core. This is because such games are already defined as *strongly balanced games* (see e.g. [17]).

6 Concluding Remarks

Selections of an interval game are very important, since they do not contain any additional uncertainty. On the top of that, selection-based classes and strong core and imputation have crucial property that although we deal with uncertain data, all possible outcomes preserve important properties. In case of selection classes it is preserving superadditivity, supermodularity etc. In case of strong core it is an invariant of having particular stable payoffs in each selection. Furthermore, "weak" concepts like $\mathcal{SC}$ are important as well since if $\mathcal{SC}$ is empty, no selection has stable payoff.

The importance of studying selection-based classes instead of the existing classes using $\succeq$ operator can be further illustrated by the following two facts:

– Classes based on weakly better operator may contain games with selections that do not have any link with the properties of their border games and consequently no link with the name of the class. For example, superadditive interval games may contain a selection that is not superadditive.
– Selection-based classes are not contained in corresponding classes based on weakly better operator. Therefore, the results on existing classes are not directly extendable to selection-based classes.

Our results provide an important tool for handling cooperative situations involving interval uncertainty which is a very common situation in various OR problems. Some specific applications of interval games were already examined. See [18–20] for applications to mountain situations, airport games and forest situations, respectively. However, these papers do not use a selection-based approach and therefore to study implications of our approach to them can be a theme for another research.

To further study properties of selection-based classes is a possible topic. One of the directions could be to introduce strictly selection convex games or decomposable games and examine them. Fruitful direction can be an extension of the definition of stable set to interval games in a selection way. For example, one could look on a union or an intersection of stable sets for each selection. Studying Shapley value and other concepts in interval games context may be interesting as well. Some of these issues are work in progress.

Acknowledgements. The authors were supported by the Czech Science Foundation Grant P402/13-10660S. The work was supported by the grant SVV 2015 260223.

References

1. Branzei, R., Dimitrov, D., Pickl, S., Tijs, S.: How to cope with division problems under interval uncertainty of claims? Int. J. Uncertainty, Fuzziness Knowl. Based Syst. **12**, 191–200 (2004)
2. Alparslan Gök, S.Z.: Cooperative interval games. Ph.D. thesis. Middle East Technical University (2009)
3. Branzei, R., Branzei, O., Alparslan Gök, S.Z., Tijs, S.: Cooperative interval games: a survey. CEJOR **18**, 397–411 (2010)
4. Branzei, R., Dimitrov, D., Tijs, S.: Models in Cooperative Game Theory, vol. 556. Springer, Heidelberg (2000)
5. Driessen, T.: Cooperative Games, Solutions and Applications, vol. 66. Kluwer Academic Publishers, Dordrecht (1988)
6. Gilles, R.P.: The Cooperative Game Theory of Networks and Hierarchies. Theory and Decision. Springer, Heidelberg (2010)
7. Peleg, B., Sudhölter, P.: Introduction to the Theory of Cooperative Games, vol. 34. Springer, Heidelberg (2007)
8. Bilbao, J.M.: Cooperative Games on Combinatorial Structures. Kluwer Academic, Boston (2000)
9. Curiel, I.: Cooperative Game Theory and Applications: Cooperative Games Arising from Combinatorial Optimization Problems, vol. 16. Springer, Heidelberg (1997)
10. Lemaire, J.: Cooperative game theory and its insurance applications. Wharton School of the University of Pennsylvania, Center for Research on Risk and Insurance (1991)
11. Caulier, J.F.: A note on the monotonicity and superadditivity of TU cooperative games. University of Tampere CEREC Working Paper (2009)
12. Shapley, L.S.: Cores of convex games. Int. J. Game Theor. **1**, 11–26 (1971)
13. Moore, R.E., Kearfott, R.B., Cloud, M.J.: Introduction to Interval Analysis. Society for Industrial and Applied Mathematics (SIAM), Philadelphia (2009)
14. Puerto, J., Fernández, F.R., Hinojosa, Y.: Partially ordered cooperative games: extended core and shapley value. Ann. Oper. Res. **158**, 143–159 (2008)
15. Alparslan Gök, S.Z., Branzei, R., Tijs, S.: Convex interval games. Advances in Decision Sciences 2009 (2009)
16. Alparslan Gök, S., Branzei, O., Branzei, R., Tijs, S.: Set-valued solution concepts using interval-type payoffs for interval games. J. Math. Econ. **47**, 621–626 (2011)
17. Alparslan Gök, S.Z., Branzei, R., Tijs, S.: Cores and stable sets for interval-valued games. Tilburg University Working Paper (2008)
18. Alparslan Gök, S.Z., Palancı, O., Olgun, M.: Cooperative interval games: mountain situations with interval data. J. Comput. Appl. Math. **259**, 622–632 (2014)
19. Alparslan Gök, S.Z., Branzei, R., Tijs, S.: Airport interval games and their shapley value. Oper. Res. Decisions **2**, 9–18 (2009)
20. Alparslan Gök, S.Z., Palanci, O., Weber, G.W.: Cooperative interval games: Forest situations with interval data. In: GAME THEORY AND MANAGEMENT. Collected abstracts of papers presented on the Seventh International Conference Game Theory and Management/Editors Leon A. Petrosyan and Nikolay A. Zenkevich.-SPb.: Graduate School of Management SPbU, 2013.-274 p. Volume 26. (2013) 180

Re-aggregation Heuristic for Large P-median Problems

Matej Cebecauer[(⊠)] and Ľuboš Buzna

University of Zilina, Univerzitna 8215/1, 01026 Zilina, Slovakia
matej.cebecauer@gmail.com, lubos.buzna@fri.uniza.sk

Abstract. When a location problem is too large to be computed in a reasonable time or if it is impossible to store it in the computer memory, an aggregation is commonly used tool that allows for transforming it to smaller size. Typically, an aggregation method is used only once, in the initial phase, before the solving process. An unavoidable consequence of the aggregation is the loss of optimality due to aggregation errors. The effects of aggregation errors might be significant, especially, when solving spatially large problems with huge number of potential customers. Here, we propose new simple re-aggregation approach, which minimizes sources of the aggregation errors. Our method aggregates the original problem to the size that can be solved by the used optimization algorithm, and in each iteration, the aggregated problem is adapted to achieve more precise location of facilities. The proposed approach is versatile and thus it can be easily extended to other location problems. To investigate the optimality error, we use benchmarks that can be computed exactly and to test the applicability of our approach, we study large benchmarks reaching 670 000 customers. Numerical experiments reveal that the re-aggregation heuristic works well for commonly used, as well as for extremely large sizes of the p-median problem. The approach allows for finding solutions of higher quality than the exact method that is applied to the aggregated problem.

Keywords: P-median problem · Re-aggregation heuristic · Open-StreetMap

1 Introduction

The location problem consists of finding a suitable set of facility locations from where services could be efficiently distributed to customers [8–10]. Many location problems are known to be NP-hard. Consequently, the ability of algorithms to compute the optimal solution quickly decreases as the problem size is growing. There are two basic approaches how to deal with this difficulty. First approach is to use a heuristic method, which, does not guarantee that we find the optimal solution. Second approach is to use the aggregation, that lowers the number of customers and candidate locations. The aggregated location problem (ALP) can be solved by an exact method or a heuristic. Unfortunately, aggregation induces

© Springer International Publishing Switzerland 2015
D. de Werra et al. (Eds.): ICORES 2015, CCIS 577, pp. 54–70, 2015.
DOI: 10.1007/978-3-319-27680-9_4

various types of errors. There is a strong stream of literature studying aggregation methods and corresponding errors [12,14]. For example, various sources of aggregation errors and approaches to minimize them are discussed by [7,12,19].

Here, we are specifically interested in finding the efficient design of the public service system that is serving spatially large geographical area with many customers. Customers are modelled by a set of demand points (DP) representing their spatial locations [14]. To include all possible locations of customers as DPs is often impossible and also unnecessary.

The basic data requirements for public service system design problem are location of DPs and the road infrastructure that is used to distribute services or access the service centres. We use volunteered geographical information (VGI) to extract road infrastructure and locations of customers. VGI is created by volunteers, who produce data through Web 2.0 applications and combine it with the publicly available data [16]. We use data extracted from the OpenStreetMap (OSM), that is one of the most successful examples of VGI. For instance, in Germany the OSM data are becoming comparable in the quality to commercial providers [28]. Road and street networks in the UK reach very good precision as well [18]. Therefore, OSM is becoming an interesting and freely available alternative to commercial datasets. We combine the OpenStreetMap data with available residential population grid [3] to estimate the demand that is associated to DPs.

It is well-known, that the solution provided by a heuristic method using more detailed data is often better than a solution achieved by the exact method, when solving the aggregated problem [1,22]. Typically, aggregation is used only in the initial phase of the solving process, to match the problem size with the performance of the used solving method. In this paper, we propose re-aggregation heuristic, where the solved problem is in each iteration modified to minimize the aggregation error in the following iterations. Our results show that the re-aggregation heuristic may provide better solutions than the exact method, which uses aggregation only once.

The paper is organized as follows: in Sect. 2.1, we briefly summarize the p-median problem, that we selected as a test case. Section 2.2 introduces the data processing procedure. The re-aggregation heuristic is explained in Sect. 2.3. Results of numerical experiments are reported in Sect. 3. We conclude in Sect. 4.

2 Materials and Methods

2.1 The P-median Problem

The number of existing location problems is overwhelming [8–10]. To evaluate the optimality error and the time efficiency of the proposed re-aggregation algorithm, we use the p-median problem, which is one of the most frequently studied and used location problems [4,6,17,23]. This problem includes all basic decisions involved in the service system design. The goal is to locate exactly p facilities in a way that the sum of weighted distances from all customers to their closest facilities is minimized. The problem is NP hard [24]. For comprehensive

overview of applications and solving methods see [25, 26]. Exact solving methods
are summarized in [30] and heuristic methods in [27].

To describe the p-median problem, we adopt the well-known integer formulation proposed in [31]. As possible candidate locations we consider all n DPs.
The length of the shortest path on the network between DP i and j is denoted
as d_{ij}. We associate to each DP the weight w_i, representing the number of customers assigned to the DP i. The decisions to be made are described by the set
of binary variables:

$$x_{ij} = \begin{cases} 1, & \text{if demand point } i \text{ is assigned to facility } j \\ 0, & \text{otherwise,} \end{cases} \tag{1}$$

$$y_j = \begin{cases} 1, & \text{if a facility at the candidate location} j \text{ is open} \\ 0, & \text{otherwise.} \end{cases} \tag{2}$$

The p-median problem can be formulated as follows:

$$\text{Minimize} \qquad f = \sum_{i=1}^{n} \sum_{j=1}^{n} w_i d_{ij} x_{ij} \tag{3}$$

subject to

$$\sum_{j=1}^{n} x_{ij} = 1 \qquad \text{for all } i = 1, 2, \ldots, n \tag{4}$$

$$x_{ij} \leq y_j \qquad \text{for all } i, j = 1, 2, \ldots, n \tag{5}$$

$$\sum_{j=1}^{n} y_j = p \tag{6}$$

$$x_{ij}, y_j \in \{0, 1\} \qquad \text{for all } i, j = 1, 2, \ldots, n. \tag{7}$$

Objective function (3) minimizes the sum of weighted distances from all DPs
to the their closest facilities. The constraints (4) insure that each customer is
allocated exactly to one facility. The constraints (5) allow for allocating customers only to located facilities and the constraint (6) makes sure that exactly
p facilities are located.

2.2   Data Model

The OSM provides all necessary data to generate DP locations and to extract
the road network. To estimate the position of demand points, we use OSM layers
describing positions of buildings, roads, residential, industrial and commercial
areas. To generate DPs, we use simple procedure. First, we generate spatial grid,
which consists of uniform square cells with the size of 100 meters. For each cell,
we extract from OSM layers elements that are situated inside the cell. Second,

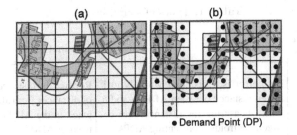

Fig. 1. Schematic illustrating the generation of DPs.

DPs are located as centroids of cells with a non empty content. The process of generating DPs is visualized in Fig. 1. Third, generated DPs are connected to the road network and we compute shortest paths between them. Finally, we calculate Voronoi diagrams, while using DP as generating points, and we associate with each DP a demand by intersecting Voronoi polygons with residential population grids produced by [3].

2.3 Aggregation Errors

Our approach is to re-aggregate the solved problem in each iteration, to achieve more precise locations of facilities in the following iterations. Aggregation is an essential part of the heuristic and it may lead to locations errors [12,14]. To minimize the effect of aggregation errors, we need to understand the possible sources of errors. Therefore, we start by a brief overview of known sources of aggregation errors that are related to the input data. Sources of errors are in the literature denoted as A, B, C and D. We describe methods how to reduce them [7,12,20,21] and to supplement this discussion, we also point at the sources of errors that are often made by designers of public systems [12].

Aggregation errors are caused by the loss of information, when DPs are replaced by aggregated demand points (ADP). Authors in [19] named these errors as source errors and introduced source A, B and C errors. Elimination of source A and B errors was studied by [7]. Minimization of the source C error was analysed in [20]. Source D error and possibilities how it can be minimized were studied in [21]. We summarize types of source errors in Table 1.

Some errors are often made by designers or decision makers, who are responsible for preparing the input data or evaluating the aggregation errors. Examples of such errors are use of the uniform demand distribution, aggregation method that is ignoring population clusters, or incorrect methods used to measure the aggregation error [12].

Proposed heuristic is able to minimize all types of source errors (A, B, C and D).

Table 1. Types of source errors.

	Source error	Description	Elimination
Eliminated by pre-processing the input data	A	This error is a result of wrongly esti-mated distance between ADPs a and b, when measuring the distance only be-tween centroids representing ADPs.	Replace the distance by the sum of distances from all DPs aggre-gated in the ADP a to the cen-troid of ADP b.
	B	It is a specific case of source A error. If ADP a is candidate location for a facil-ity, and at the same time it represents a customer, the distance between facility location a and customer a is incorrectly set to zero value.	Replace the zero distance by the sum of all distances from DPs aggregated in the ADP a to the centroid of the ADP a.
Eliminated by post-processing the input data	C	All DPs aggregated in the same ADP are assigned to the same facility.	Re-aggregate ADPs and find the closest facility for all DPs.
	D	Facility is established in ADP centroid and not in DP, thus the location of facil-ity is almost certainly not optimal with respect to its service zone.	Find DP, where facility is lo-cated, by disaggregating ADPs in the close neighbourhood of the located facility.

2.4   The Row-Column Aggregation Method

To aggregate DP, we use row-column aggregation method proposed in [1,13]. We review the original row-column aggregation method and describe its adaptation to the spatially large geographical areas with many municipalities.

First, we list three basics steps of the original aggregation method [1,13]:

Step 1: Generate irregular grid for the whole geographical area of the problem.
Step 2: Select centroid of each grid cell representing ADP.
Step 3: Assign each DP to the closest ADP.

The irregular grid, with c columns and r rows, is obtained by sequentially solving the c-median problem on the projection of the DPs to the x-axis, and by solving the r-median problem on the projection of the DPs to the y-axis. The border lines defining the rows and columns are positioned in the middle between facili-ties that has been found by solving the one dimensional location problems [13].

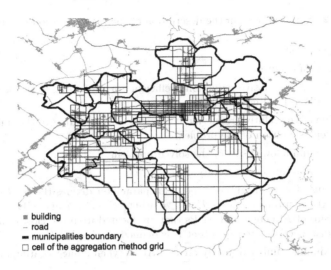

Fig. 2. Map of the area after the application of the row-column aggregation method to each administrative zone separately. Rectangles corresponds to grid cells.

Next in the step 2, for each cell of the grid, we extract the subnetwork of the road network that intersects with the area of the cell. ADP is found by solving the 1-median problem for each individual subnetwork [1]. Finally, each DP is assigned to the closest ADP.

We apply this approach to each individual municipality separately. This allows for approximating the population clusters more precisely and thus it helps to minimize the aggregation error. In the Fig. 2 is visualised a result of row-column aggregation method.

2.5   The Re-aggregation Heuristic

In order to characterize the size of the aggregated problem, we define the relative reduction coefficient:

$$\alpha = (1 - \frac{number\ of\ ADPs}{number\ of\ original\ DPs})100\,\%. \qquad (8)$$

Thus, for the unaggregated problem we get the value $\alpha = 0$. Used notation is described in Table 2.

We propose the re-aggregation heuristic composed from seven phases that are executed in the following order:

Phase 0: Initialization
Set $i = 1$, prepare ALP, with the size corresponding to the reduction coefficient $\alpha(1)$, by aggregating the input data and compute the distance matrix.
Phase 1: Elimination of Source A and B Errors
Update the distance matrix accounting for source A and B Errors.

Table 2. Notation used in the description of the re-aggregation heuristic.

Symbol	Description
i	Iteration counter
$\alpha(i)$	Value of the relative reduction coefficient corresponding to the solved ALP in the iteration i
$\alpha(1)$	Value of the relative reduction coefficient corresponding to the initial representation of the problem
α_{max}	Maximum allowed size of the solved problem
i_{max}	Maximum number of iterations
ϵ	Radius of ADP neighbourhood. This parameter divides the set I of all ADPs into two subsets $E, F \subset I$, where $E \cap F = \emptyset$ and $E \cup F = I$. Subset E includes all ADPs that are located from the closest facility at distance less than ϵ. Subset F is defined as $F = I - E$
λ	Maximum number of newly created ADPs when de-aggregating an ADP

Phase 2: Location of Facilities
Solve the p-median problem. As a result we obtain p located facilities.

Phase 3: Elimination of Source C and D Errors
Minimize the source C error by reallocating DPs to the closest facilities.
Minimize the source D error by decomposing the problem into p location problems, each consisting of one located facility and of all associated DPs. For each decomposed problem solve the 1-median problem. As a result we obtain p newly located facilities.

Phase 4: Identification of ADPs located in the Central Area of Service Zones
Set $E = \emptyset$ and $F = \emptyset$. Process all ADPs and if distance from ADP to the closest facility is less than ϵ then insert ADP into the set E, otherwise insert ADP into the set F.

Phase 5: Identification of ADPs located in the Boundary Area of Service Zones
Move from the set F into the set E all ADPs that include at least one DP that has shorter distance to another facility than its ADP centroid.

Phase 6: Re-aggregation
If every facility is established in a DP that cannot be further de-aggregated or if $i > i_{max}$ then terminate. Output the best found solution as the final result. Otherwise, increment i by 1 and de-aggregate each ADP in the set E to at maximum λ new ADPs, using the row-column aggregation method and update value $\alpha(i)$. If $\alpha(i) > \alpha_{max}$ then terminate, otherwise go to phase 1.

2.6   Evaluation of Performance

By means of numerical experiments, we investigate the relation between quality of the solution and computational time. Adopting definitions described in [11], we formalize the relative error Δ as:

$$\Delta(x_\alpha, y) = \frac{f(y) - f(x_\alpha)}{f(x_\alpha)} 100\,\%, \tag{9}$$

where $f()$ is the value of the objective function (3), y is the solution provided by our re-aggregation heuristic and x_α is the optimal solution of the ALP problem with the relative reduction α. Thus, x_0 denotes the optimal solution of the original, i.e. unaggregated, problem. When we use x_0 in the formula (9), we recover the optimality error defined in [11].

Using the same notation, we define the relative time efficiency τ as:

$$\tau(x_\alpha, y) = \frac{t(y) - t(x_\alpha)}{t(x_\alpha)} 100\,\%, \tag{10}$$

where $t()$ is the time spent by computing the solution.

3   Numerical Results

We apply re-aggregation heuristic to four geographical areas. More details about selected geographical areas are given in Table 3. To solve the p-median problem, we use the state-of-the-art algorithm ZEBRA [15]. We evaluate the importance of individual phases of the proposed heuristic by formulating and comparing four different versions of the re-aggregation heuristic. We denote them as $V1$, $V2$, $V3$ and $V4$. Table 4 summarizes the presence of individual phases in each version.

Table 3. Geographical areas that constitute our benchmarks.

Area	Number of DPS	Size[km^2]	Population
Partizánske	4,873	301	47,801
Košice	9,562	240	235,251
Žilina	79,612	6,809	690,420
Slovakia	663,203	49,035	5,418,561

To save some computational time, in the version $V1$ we skip the phase 5. As we will show later, this modification is especially suitable for very large location problems, where the re-aggregation heuristic is very time consuming otherwise.

We organized numerical experiments into two groups. We start investigating the performance of the re-aggregation heuristic by studying benchmarks Partizánske and Košice. These two benchmarks are small enough to be computed to optimality and thus we can compare our results with the exact algorithm. Moreover, we solve large benchmarks Žilina and Slovakia. These two benchmarks are too large to be solved to optimality.

Table 4. Phases included in each version of the re-aggregation heuristic. By the symbol "+" we indicate the presence of the phase.

Version	Phases						
	0	1	2	3	4	5	6
V1	+		+		+		+
V2	+		+		+	+	+
V3	+	+	+		+	+	+
V4	+	+	+	+	+	+	+

Table 5. Results of numerical experiments for the geographical area of Partizánske. After the symbol "|" in α-row, we detail the number of iterations i reached when the algorithm terminated.

	ϵ	$\alpha(1)$	p=5				p=10				p=20			
	[km]	[%]	V1	V2	V3	V4	V1	V2	V3	V4	V1	V2	V3	V4
$\alpha(i)[\%]\|i$	0	1	97.9|7	95.3|9	95.6|10	94.8|9	96.3|9	91.0|12	89.6|17	89.8|11	93.1|11	86.2|14	84.6|13	84.1|10
$\Delta[\%]$	0	1	1.237	0.772	0.933	0.173	3.519	1.831	1.465	0	3.654	1.566	1.224	0.105
$\tau[\%]$	0	1	−99.9	−99.6	−99.1	−97.5	−99.6	−96.4	−91.3	−93.5	−97.2	−71.7	−57.2	−63.7
$\alpha(i)[\%]\|i$	0	10	90.5|4	89.3|6	89.4|6	89.1|5	86.8|8	86.9|7	87.2|7	87.1|6	88.6|5	83.1|12	83.8|11	83.3|8
$\Delta[\%]$	0	10	0.477	0.200	0.200	0.173	0.934	0.913	0.891	0.365	2.619	1.187	1.080	0.017
$\tau[\%]$	0	10	−99.4	−98.6	−96.7	−96.3	−96.2	−94.7	−93.7	−92.9	−96.5	−64.4	−53.8	−61.1
$\alpha(i)[\%]\|i$	0	25	77.6|3	77.0|5	77.1|4	77.1|5	77.2|5	75.6|8	75.1|8	75.7|5	76.5|4	74.8|6	74.0|8	73.7|6
$\Delta[\%]$	0	25	0.037	0.037	0.037	0	0.644	0.529	0.308	0.043	0.883	1.140	0.856	0.017
$\tau[\%]$	0	25	−97.7	−95.8	−86.8	−81.4	−95.7	−90.1	−78.6	−85.8	−87.6	−72.3	−39.3	−54.4
$\alpha(i)[\%]\|i$	1	1	85.9|10	77.9|14	78.4|12	79.3|12	62.9|13	62.9|13	59.9|14	62.9|12	53.4|13	42.3|12	45.0|13	46.8|12
$\Delta[\%]$	1	1	0.567	0.235	0.173	0.173	0.087	0.087	0		0.393	0.115	0.292	0.105
$\tau[\%]$	1	1	−99.2	−93.0	−80.6	−80.2	−81	−67.1	−35.3	−55.1	−21.4	125.1	221.0	153.8
$\alpha(i)[\%]\|i$	1	10	75.4|7	74.0|11	73.3|11	75.2|11	62.8|11	62.8|11	62.7|11	62.5|11	44.8|8	43.6|11	43.7|11	44.8|11
$\Delta[\%]$	1	10	0.173	0.173	0.173	0.173	0.151	0.151	0.148	0	0.244	0.116	0.105	0.105
$\tau[\%]$	1	10	−96.7	−87.9	−52.6	−61.4	−75.3	−35.4	−35.1	−25.9	−35.4	156.4	202.9	223.6
$\alpha(i)[\%]\|i$	1	25	67.8|5	65.9|10	66.1|10	66.3|10	58.6|6	55.6|10	55.5|10	56.6|10	44.8|6	41.1|10	41.1|10	41.4|10
$\Delta[\%]$	1	25	0	0	0	0	0.003	0.003	0	0	0.049	0.049	0.031	0
$\tau[\%]$	1	25	−93.8	−82.5	−29.0	−37.3	−88.2	−65.3	−32.1	−33.7	−34.6	66.0	116.3	133.4
$\alpha(i)[\%]\|i$	2	1	66.4|10	59.3|15	59.9|15	60.0|15	45.8|10	36.0|15	37.0|15	38.8|15	20.7|10	17.5|15	17.3|15	18.1|15
$\Delta[\%]$	2	1	0.040	0.026	0	0	0.208	0	0	0	0.001	0.001	0	0
$\tau[\%]$	2	1	−93.1	−75.5	−24.3	−28.3	−67.7	19.4	97.1	53.5	103.5	474.9	527.3	509.0
$\alpha(i)[\%]\|i$	2	10	58.7|7	56.8|12	53.5|10	57.2|11	40.7|7	37.4|11	37.6|11	37.8|11	20.5|8	17.5|11	17.3|11	17.4|11
$\Delta[\%]$	2	10	0.015	0	0	0	0	0	0	0	0	0.001	0	0
$\tau[\%]$	2	10	−88.7	−64.3	−13.0	−0.5	−67.6	−1.9	58.7	84.2	103.1	289.8	351.7	349.9
$\alpha(i)[\%]\|i$	2	25	54.8|5	53.3|10	53.5|10	53.5|10	37.8|6	35.7|10	35.9|10	36.1|10	19.1|8	17.3|10	16.9|10	17.8|10
$\Delta[\%]$	2	25	0	0	0	0	0.003	0	0	0	0.003	0.001	0	0
$\tau[\%]$	2	25	−86.3	−63.6	−13.0	−9.6	−70.5	−14.7	119.0	39.2	155.8	278.6	330.6	337.7

3.1 Benchmarks Partizánske and Košice

We investigate the impact of parameters ϵ and $\alpha(1)$ on the optimality error Δ and time efficiency τ. We compare three different values of the input parameter $\alpha(1)$: $1\%, 10\%$ and 25%. Further, we investigate three values of the parameter ϵ: 0, i.e. the surrounding of facilities is not re-aggregated and values $1\,km$ and $2\,km$ when all ADPs closer than 1 or 2 kilometres from located facilities are re-aggregated. We set α_{max} to value 0%.

Table 6. Results of numerical experiments for the geographical area of Košice. After the symbol "|" in α-row, we detail the number of iterations i reached when the algorithm terminated.

	ϵ [km]	$\alpha(1)$ [%]	p=10				p=20				p=40			
			V1	V2	V3	V4	V1	V2	V3	V4	V1	V2	V3	V4
$\alpha(i)$[%]|i	0	1	97.8|7	91.8|13	91.7|20	91.6|19	96.3|11	89.0|9	87.3|16	87.2|14	92.7|14	82.3|21	83.5|21	84.0|16
Δ[%]	0	1	4.025	2.155	1.699	0.099	2.539	2.020	1.259	0.177	3.252	1.177	0.480	0.018
τ[%]	0	1	−99.9	−98.4	−93.4	−91.1	−99.8	−98.3	−90.6	−91.5	−97.5	−66.5	−53.6	−63.2
$\alpha(i)$[%]|i	0	10	87.6|12	87.5|12	87.6|12	86.8|12	89.0|6	84.6|9	84.7|9	85.5|8	88.0|6	81.8|12	81.1|14	80.4|16
Δ[%]	0	10	1.167	1.167	0.906	0.005	1.592	0.801	1.030	0	2.364	2.151	1.399	0.035
τ[%]	0	10	−96.8	−96.7	−91.3	−89.2	−99.1	−96.3	−92.6	−93.7	−97.2	−78.5	−54.6	−44.5
$\alpha(i)$[%]|i	0	25	74.9|4	73.9|6	74.0|7	73.9|7	74.6|5	72.5|6	72.7|7	72.4|7	74.1|4	71.0|10	71.8|9	70.7|14
Δ[%]	0	25	0.093	0.033	0.037	0	1.004	0.740	0.890	0	1.298	1.198	1.156	0.191
τ[%]	0	25	−96.5	−94.7	−81.9	−81.0	−93.2	−92.2	−79.7	−82.1	−88.8	−63.1	−46.2	−12.2
$\alpha(i)$[%]|i	1	1	71.2|16	71.2|16	70.9|16	70.9|14	63.2|11	57.4|16	57.4|13	57.6|13	50.4|11	44.5|17	45.3|18	44.7|17
Δ[%]	1	1	0.008	0.008	0	0	0.359	0.186	0.035	0.036	0.089	0.023	0.017	0.014
τ[%]	1	1	−85.7	−83.3	−61.1	−64.6	−84.7	−56.7	−44.0	−45.5	−27.7	106.7	215.0	196.8
$\alpha(i)$[%]|i	1	10	68.8|12	68.9|12	68.9|12	69.3|12	58.9|9	55.4|12	57.4|12	57.8|12	47.9|8	43.0|18	43.5|16	43.7|17
Δ[%]	1	10	0.004	0.004	0	0	0.229	0.190	0.150	0.150	0.079	0.020	0.017	0
τ[%]	1	10	−84.7	−78.3	−54.3	−51.2	−72.7	−51.0	−28.1	−29.0	−32.0	239.3	260.8	342.9
$\alpha(i)$[%]|i	1	25	63.8|6	62.0|8	61.9|8	61.9|8	53.5|7	50.9|13	51.5|13	52.1|13	44.21|7	41.5|14	41.8|14	42.2|12
Δ[%]	1	25	0.009	0.009	0	0	0.003	0.003	0	0	0.024	0.020	0.014	0.014
τ[%]	1	25	−86.1	−82.6	−53.4	−57.9	−65.6	−39.9	−14.0	−11.5	−25.7	102.3	183.6	127.4
$\alpha(i)$[%]|i	2	1	47.9|11	44.7|19	44.6|19	44.8|16	33.9|10	31.1|19	29.5|20	31.4|18	24.2|10	20.2|18	20.3|18	22.1|18
Δ[%]	2	1	0.008	0	0	0	0.039	0.039	0	0	0.032	0	0	0
τ[%]	2	1	−26.1	−24.1	62.1	23.7	−25.3	79.0	168.2	110.0	146.3	473.4	630.2	563.6
$\alpha(i)$[%]|i	2	10	46.8|10	44.0|12	44.1|12	44.1|12	33.0|10	29.3|18	29.6|18	29.7|16	23.1|10	21.0|15	20.9|15	21.6|15
Δ[%]	2	10	0	0	0	0	0.036	0	0	0	0.011	0	0	0
τ[%]	2	10	−54.3	−24.4	1.4	18.3	−13.9	109.4	124.7	99.6	187.7	392.7	491.2	459.5
$\alpha(i)$[%]|i	2	25	42.2|7	40.5|17	41.1|17	41.1|17	29.5|8	27.9|17	28.0|17	27.8|18	21.0|10	19.2|17	19.6|17	20.0|17
Δ[%]	2	25	0	0	0	0	0.004	0.004	0	0	0.004	0	0	0
τ[%]	2	25	−54.4	45.1	130.1	151.0	−13.9	98.0	156.2	171.8	187.7	555.7	855.2	699.6

The results of numerical experiments are reported in Tables 5 and 6. In the majority of cases, we find the optimality error Δ smaller than 1 %. For the area of Partizánske and Košice, when $\epsilon = 0$, we find few cases when the value of the optimality error Δ is larger, but here the reduction coefficient α is very high, which means that much lower number of ADPs was taken into account.

By varying ϵ, we are trading the optimality error for the computational time. The results in Tables 5 and 6 show how the likelihood of finding the optimal solutions is growing when ϵ is increasing. The likelihood of finding the optimal solution for the $\epsilon = 0$ is 2/36 for the benchmark Partizánske and 3/36 for the benchmark Košice, while for the $\epsilon = 1$ it increased to 10/36 for the benchmark Partizánske and to 9/36 for Košice. When $\epsilon = 2$, in the majority of cases we find the optimal solution (26 out of 36 for Partizánske and 27 out of 36 for Košice). The time efficiency decreases with growing the parameter ϵ, because re-aggregations in the surrounding of facilities decrease the reduction coefficient.

Typically, the larger is the value of parameter $\alpha(1)$, the better is the quality of the solution. As the parameter $\alpha(1)$ grows the computational time has tendency to grow, as well. However, this tendency is reversed in some cases.

Version $V4$ provides solutions with the best quality. The optimal solution was always found when $\epsilon = 2$, in 55 % of cases when $\epsilon = 1$ and in 22 % of cases when $\epsilon = 0$. When ordering the versions, with respect to the frequency of finding the optimal solution, we obtain the following order: V4,V3,V2 and V1. Version $V1$ provides no optimal solution with $\epsilon = 0$, in $= 0.05$ % of cases when $\epsilon = 1$ and only in 28 % of cases when $\epsilon = 2$. Version $V1$ is the most time efficient. It reaches the lowest number of iterations and the highest final value of the reduction coefficient α. The most time consuming are versions $V3$ and $V4$. Both of them include phase 1. Thus, we can conclude that the elimination of source A and B errors is largely time consuming. Unexpected finding is that the elimination of source A and B errors may worsen the quality of the solution.

The re-aggregation heuristic has higher relative time efficiency τ, when p is small. This can be particularly beneficial. For example the state-of-the-art-algorithm ZEBRA systematically consumes more computational time and more computer memory when p is small. The computer memory allocation needed to find optimal solution for the benchmark Košice using the algorithm ZEBRA for $p = 5$ is 10.41 GB. Our re-aggregation heuristic demanded less than 3 GB. The relative time efficiency τ decreases when increasing the value of p. This can be explained by small value of the reduction coefficient α when p is larger.

In the majority of cases, re-aggregation heuristic has much higher time efficiency than the exact algorithm (value τ is negative). Typically, the heuristic terminates when no change in the ALP occurs. It is important to note, that typically all versions of the heuristic encountered the best solution after few iterations and in the following iterations the solution did not change. As an example, we show the behaviour of the re-aggregation algorithm detailing individual iterations (see Table 7).

Table 7. Typical behaviour of the re-aggregation heuristic. We evaluate each individual iteration for benchmark Partizánske and $p = 20$, $\alpha(1) = 99$ %, $\epsilon = 2$ km, $\alpha_{max} = 0$ %.

Iteration i	V1			V2			V3			V4		
	$\alpha(i)$ [%]	Δ [%]	τ [%]	$\alpha(i)$ [%]	Δ [%]	τ [%]	$\alpha(i)$ [%]	Δ [%]	τ [%]	$\alpha(i)$ [%]	Δ [%]	τ [%]
1	99.0	15.071	−99.2	99.0	15.071	−99.0	99.0	15.003	−99.0	99.0	4.124	−98.6
2	97.1	13.292	−99.0	96.9	12.761	−98.6	97.0	13.410	−98.2	96.8	2.558	−97.4
3	90.1	5.574	−98.4	89.8	4.983	−96.9	90.2	4.858	−96.1	89.3	0.133	−94.7
4	70.8	1.750	−94.5	70.7	2.026	−89.4	71.7	2.694	−87.2	68.9	0.105	−83.9
5	42.8	0.282	−78.2	45.1	0.246	−65.8	46.1	0.310	−59.7	42.9	0.115	−55.4
6	27.9	0.084	−50.3	28.7	0.088	−23.0	29.1	0.265	−17.3	27.2	0	−12.4
7	22.8	0.018	−15.3	21.8	0.018	33.0	21.9	0.017	36.0	21.6	0	42.4
8	21.0	0.001	31.6	19.1	0.001	90.6	19.0	0	91.4	19.4	0	100.4
9	20.7	0.001	67.6	18.1	0.001	150.9	18.0	0	152.5	18.6	0	159.7
10	20.7	0.001	103.5	17.7	0.001	215.5	17.6	0	214.9	18.3	0	219.6
11	-	-	-	17.6	0.001	278.2	17.5	0	281.5	18.2	0	277.6
12	-	-	-	17.5	0.001	330.6	17.4	0	342.2	18.1	0	337.9
13	-	-	-	17.5	0.001	377.0	17.4	0	440.9	18.1	0	394.9
14	-	-	-	17.5	0.001	423.4	17.4	0	469.7	18.1	0	451.9
15	-	-	-	17.5	0.001	474.9	17.3	0	527.3	18.1	0	509.0

In the next subsection, we present the results obtained on large benchmarks Žilina and Slovakia. In contrast to small problems, we compute the shortest path distances on the fly. This has to be done, when the size of the problem does not allow to store the distance matrix in the computer memory. This leads to larger computations times and makes impossible the comparison of computational times between small and large problem instances.

3.2 Benchmarks Žilina and Slovakia

We compute large location problem Žilina using V2,V3 and V4 versions of the re-aggregation algorithm. The parameter $\alpha(1)$ we keep fixed to value 99 % and the parameter α_{max} to value 90 %. We investigate the optimality error and the computational time. We evaluate only first three iterations of the algorithm, because the size of the ADP rapidly grows. Consequently, the computational time becomes quickly very long.

When designing a public service system, it is common in location analyses to aggregate DPs to the level of municipalities [23]. For the region of Žilina, we also prepared such aggregated problem. Here we obtained 346 ADPs, corresponding to $\alpha_1 = 99.6\%$. The size of the benchmark Žilina does not allow to compute the optimal solution, and thus, we cannot evaluate the optimality error. Therefore, using the row-column aggregation method, we prepared benchmark Žilina with the value of $\alpha_2 = 90\%$.

Table 8. Relative error $\Delta(x_\alpha, y)$ for benchmark Žilina, evaluated after each individual iteration ($\alpha(1) = 99.0\%$ and $\alpha_{max} = 90\%$).

Version	Iteration i	$\alpha(i)$ [%]	Number of ADPs	t[h]	$\Delta(x_\alpha, y)$ [%]	
					346 ADPs($\alpha_1 = 99.6$)	7 961 ADPs ($\alpha_2 = 90$)
V2	1	99.0	796	0.3	−1.1	1.4
	2	97.4	2 062	1.3	−2.4	0.19
	3	94.4	4 424	8.5	−2.7	−0.01
V3	1	99.0	796	0.3	−1.1	1.49
	2	97.4	2 088	3.0	−2.3	0.28
	3	94.6	4 265	23.6	−2.6	−0.03
V4	1	99.0	796	33.8	−2.93	−0.38
	2	97.5	2 013	69.9	−2.93	−0.38
	3	94.8	4 146	114.7	−2.93	−0.38

The results in Table 8 show that versions $V2$ and $V3$ improve the solution in each iteration, while version $V4$ does not change the solution within this range of iterations. When the re-aggregation heuristic is initialized by ALP consisting from 796 ADPs, it enables to achieve better solution (saving in total 20 757 km), than is the optimal solution obtained with fixed size problem consisting from 7 961 ADPs. Although, the computational time is longer.

In the case of the large location problems the most time consuming phase is the elimination of the source C and D errors. Especially, the elimination of the

Table 9. Relative errors $\Delta(x_\alpha, y)$ of the re-aggregation heuristic (version V1) for the benchmark Žilina, evaluated for selected individual iterations, with respect to the optimal solutions computed on benchmarks reduced to $\alpha_1 = 99.6\,\%$ and $\alpha_2 = 90\,\%$, respectively ($\alpha(1) = 99.0\,\%$ and $\alpha_{max} = 90\,\%$).

ϵ [km]	Iteration i	$\alpha(i)$ [%]	Number of ADPs	t[h]	$\Delta(x_\alpha, y)$[%]	
					346 ADPs($\alpha_1 = 99.6$)	7 961 ADPs ($\alpha_2 = 90$)
0	1	99.00	796	0.5	−1.34	1.28
	3	98.95	837	1.4	−1.64	0.97
	5	98.94	846	1.6	−2.4	0.19
	7	98.93	851	2.2	−2.46	−0.13
1	1	99.00	796	0.5	−1.34	1.28
	3	98.65	1 074	1.0	−2.02	0.58
	5	97.87	1 695	2.1	−2.71	−0.13
	7	97.23	2 205	3.4	−2.76	−0.17
	9	97.14	2 276	4.8	−2.78	−0.20
	10	97.12	2 292	5.5	−2.78	−0.20
2	1	99.00	796	0.5	−1.34	1.28
	3	98.25	1 396	1.4	−2.52	0.07
	5	94.87	4 083	4.5	−2.93	−0.34
	7	93.26	5 362	13.0	−2.94	−0.35
	9	93.01	5 562	21.25	−3.02	−0.44
	11	93.01	5 604	31.31	−3.02	−0.44
	12	92.96	5 606	36.5	−3.02	−0.44

source D errors is very expensive, due to the frequent use of graph distances. In the version $V1$ is missing the re-aggregation of ADPs selected in the phase 5. Therefore $V1$ is much faster on large instances and it can compute more iterations. The results in Table 9 show that $V1$ is able to provide better solutions. All runs completed when all facilities are located in the ADPs that represent only one DP. Thus, the algorithm terminated because it was not possible to locate facilities more precisely using the de-aggregation.

Next, we compute extremely large benchmark Slovakia using only version $V1$. Computation with other versions of the heuristic turned out to be very time consuming. We set the parameter α_{max} to 99.3 %. Again, we compare the results with respect to exact solutions obtained for benchmarks aggregated to $\alpha_1 = 99.6\,\%$ (corresponding to 2 929 municipalities, where each municipality is represented by single ADP) and $\alpha_2 = 99.3\,\%$ (consisting from 5000 ADPs, produced by row-column aggregation method).

The computational time in Table 10 is kept within the reasonable limits and therefore we were able to run the algorithm until the solution have continued to improve. The re-aggregation heuristic provides better solution than the exact method on the fixed ALP. The version $V1$ of the re-aggregation heuristic with 3 782 ADPs and $\epsilon = 2$ reduced objective function in total by 506 604 km, using larger level of aggregation.

Table 10. Relative errors $\Delta(x_\alpha, y)$ of the re-aggregation heuristic (version V1) for the benchmark Slovakia, evaluated for selected individual iterations with respect to optimal solutions computed on the benchmarks reduced to $\alpha_1 = 99.6\,\%$ and $\alpha_2 = 99.3\,\%$, respectively ($\alpha(1) = 99.56\,\%$ and $\alpha_{max} = 99.3\,\%$).

ϵ [km]	Iteration i	$\alpha(i)$ [%]	Number of ADPs	t[h]	$\Delta(x_\alpha, y)$[%] 2 929 ADPs ($\alpha_1 = 99.6$)	5 000 ADPs ($\alpha_2 = 99.3$)
0	1	99.56	2 929	23.1	0	1.56
	3	99.55	2 981	47.2	−1.39	0.143
	5	99.54	3 034	71.9	−1.61	−0.07
	7	99.54	3 058	97.0	−1.66	−0.14
	9	99.54	3 061	121.7	−1.69	−0.15
1	1	99.56	2 929	22.1	0	1.56
	3	99.55	2 987	46.4	−1.43	0.11
	5	99.52	3 207	72.3	−1.71	−0.18
	7	99.45	3 643	103.2	−1.73	−0.19
	9	99.42	3 841	138.3	−1.74	−0.20
	10	99.42	3 842	156.1	−1.74	−0.20
2	1	99.56	2 929	22.5	0	1.56
	3	99.54	3 026	51.6	−1.6	−0.02
	5	99.43	3 782	98.0	−1.8	−0.30
	6	99.20	4 868	106.07	−1.9	−0.33

4   Conclusions

When a location problem is too large to be solved by an available method, an aggregation can help to save the situation. Typically, aggregation is done when processing the input data and it is kept separated from solving methods. In this paper, we propose and test the re-aggregation heuristic, which is adapting the granularity of the input data in each iteration of the solving process to aggregate less in areas where located facilities are situated and more elsewhere.

To determine the usability of this approach, we prepared four problems derived from geographical areas that consist of many municipalities. Please note, in the location analyses it is not common to work with such large problems. In the literature, we found only two comparable examples where the *p*-median problem with approximately 80,000 DPs is solved [2,15]. Moreover, these benchmarks were not derived from a real-world problem, but randomly generated.

Our main findings derived from experimenting with small benchmarks Partizánske and Košice are the following:

- Values of the parameter ϵ larger than 0 improve significantly the quality of the solution. However, too large values of ϵ may lead to solving exceedingly large problems and thus lead to long computational times. Therefore it is important to find reasonable trade off.

– The minimization of the source C and D errors helps to obtain better quality of the solution.
– The minimization of the source A and B errors leads to controversial effects and in some cases it deteriorates the quality of the solution.

From experimenting with large benchmarks Žilina and Slovakia we conclude:

– Elimination of the source errors is very time consuming.
– Larger values of the parameter ϵ seem to be more beneficial than elimination of the source errors.
– Among the tested variations of the algorithm, the version $V1$, where we skipped elimination of source errors and set the parameter ϵ to values 1 and 2, is the most suitable for solving very large location problems.

In this paper, we presented our initial results that are entirely based on the p-median problem. However, the proposed re-aggregation heuristic is versatile and it can be easily adjusted to other types of location problems. It is enough to replace the solving algorithms used in phases 2 and 3 with customized algorithms. More complicated location problems, such as for example the lexicographic minimax approach, have considerably larger computational complexity and the existing approaches [5, 29] do not provide solutions for problems with large number of DPs. We believe that in such cases our approach could be very promising.

Acknowledgements. This work was supported by the research grants VEGA 1/0339/13 Advanced microscopic modelling and complex data sources for designing spatially large public service systems and APVV-0760-11 Designing Fair Service Systems on Transportation Networks.

References

1. Andersson, G., Francis, R.L., Normark, T., Rayco, M.: Aggregation method experimentation for large-scale network location problems. Location Sci. **6**(1–4), 25–39 (1998)
2. Avella, P., Boccia, M., Salerno, S., Vasilyev, I.: An aggregation heuristic for large scale p-median problem. Comput. Oper. Res. **39**, 1625–1632 (2012)
3. Batista e Silva, F., Gallego, J., Lavalle, C.: A high-resolution population grid map for Europe. J. Maps 9(1), 16–28 (2013)
4. Berlin, G.N., ReVelle, C.S., Elzinga, D.J.: Determining ambulance - hospital locations for on-scene and hospital services. Environ. Plan. A **8**(5), 553–561 (1976)
5. Buzna, L., Kohani, M., Janacek, J.: An approximation algorithm for the facility location problem with lexicographic minimax objective. J. Appl. Math. **2014**(12), 1–12 (2014)
6. Calvo, A.B., Marks, D.H.: Location of health care facilities: an analytical approach. Socio-Economic Plan. Sci. **7**(5), 407–422 (1973)

7. Current, J.R., Schilling, D.A.: Elimination of source A and B errors in p-Median location problems. Geogr. Anal. **19**(8), 95–110 (1987)
8. Daskin, M.S.: Network and Discrete Location: Models, Algoritmhs and Applications. John Wiley & Sons, New York (1995)
9. Drezner, Z. (ed.): Facility location: A survey of Applications and Methods. Springer Verlag, New York (1995)
10. Eiselt, H.A., Marianov, V. (eds.): Foundations of Location Analysis. Science + Business, International Series in Operations Research and Management Science, vol. 155, Springer, US (2011)
11. Erkut, E., Neuman, S.: A multiobjective model for locating undesirable facilities. Ann. Oper. Res. **40**, 209–227 (1992)
12. Erkut, E., Bozkaya, B.: Analysis of aggregation errors for the p-median problem. Comput. Oper. Res. **26**(10–11), 1075–1096 (1999)
13. Francis, R.L., Lowe, T.J., Rayco, M.B.: Row-Column aggregation for rectilinear distance p-Median problems. Transp. Sci. **30**(2), 160–174 (1996)
14. Francis, R., Lowe, T., Rayco, M., Tamir, A.: Aggregation error for location models: survey and analysis. Ann. Oper. Res. **167**(1), 171–208 (2009)
15. García, S., Labbé, M., Marín, A.: Solving large p-Median problems with a radius formulation. INFORMS J. Comput. **23**(4), 546–556 (2011)
16. Goodchild, M.: Citizens as sensors: the world of volunteered geography. GeoJournal **69**(4), 211–221 (2007). http://dx.doi.org/10.1007/s10708-007-9111-y
17. Hakimi, S.L.: Optimum distribution of switching centers in a communication network and some related graph theoretic problems. Oper. Res. **13**(3), 462–475 (1965)
18. Haklay, M.: How good is volunteered geographical information? a comparative study of OpenStreetMap and ordnance survey datasets. Environ. Plan. B Plan. Des. **37**(4), 682–703 (2010)
19. Hillsman, E., Rhoda, R.: Errors in measuring distances from populations to service centers. Ann. Reg. Sci. **12**(3), 74–88 (1978)
20. Hodgson, M.J., Neuman, S.: A GIS approach to eliminating source C aggregation error in p-median models. Location Sci. **1**, 155–170 (1993)
21. Hodgson, M.J., Shmulevitz, F., Körkel, M.: Aggregation error effects on the discrete-space p-median model: the case of Edmonton, Canada. Can. Geogr. / Le Géographe canadien **41**(4), 415–428 (1997)
22. Hodgson, M., Hewko, J.: Aggregation and surrogation error in the p-Median model. Ann. Oper. Res. **123**(1–4), 53–66 (2003)
23. Janáček, J., Jánošíková, L., Buzna, L.: Optimized Design of Large-Scale Social Welfare Supporting Systems on Complex Networks. In: Thai, M.T., Pardalos, P.M. (eds.) Handbook of Optimization in Complex Networks, Springer Optimization and Its Applications, vol. 57, pp. 337–361. Springer, US (2012)
24. Kariv, O., Hakimi, S.L.: An algorithmic approach to network location problems. II: the p-Medians. SIAM J. Appl. Math. **37**(3), 539–560 (1979)
25. Marianov, V., Serra, D.: Location Problem in the Public Sector. In: Drezner, Z., Hamacher, H.W. (eds.) Facility Location: Applications and Theory, p. 460. Springer, Heidelberg (2002)
26. Marianov, V., Serra, D.: Median problems in networks. In: Eiselt, H.A., Marianov, V. (eds.) Foundations of Location Analysis, International Series in Operations Research & Management Science, vol. 155, pp. 39–59. Springer, US (2011)
27. Mladenović, N., Brimberg, J., Hansen, P., Moreno-Pérez, J.A.: The p-median problem: a survey of metaheuristic approaches. Eur. J. Oper. Res. **179**(3), 927–939 (2007)

28. Neis, P., Zielstra, D., Zipf, A.: The Street Network Evolution of Crowdsourced Maps: OpenStreetMap in Germany 2007–2011. Future Internet 4(1), 1–21 (2011). http://www.mdpi.com/1999-5903/4/1/1
29. Ogryczak, W.: On the lexicographic minmax approach to location problems. Eur. J. Oper. Res. **100**, 566–585 (1997)
30. Reese, J.: Solution methods for the p-median problem: an annotated bibliography. Networks **48**(3), 125–142 (2006)
31. ReVelle, C.S., Swain, R.W.: Central facilities location. Geogr. Anal. **2**(1), 30–42 (1970)

Meeting Locations in Real-Time Ridesharing Problem: A Buckets Approach

K. Aissat[1]([✉]) and A. Oulamara[2]

[1] University of Lorraine - LORIA, Nancy, France
[2] University of Lorraine, Ile de Saulcy, Metz, France
{kamel.aissat,ammar.oulamara}@loria.fr

Abstract. Improving transportation efficiency without increasing urban traffic congestion, requires to carry out new services such as ridesharing which contributes to reduce operating cost and to save road resources. In this paper, we provide users of a ridesharing system greater flexibility: given a set of drivers' offers already in the system, and a new rider's request, we determine a best driver, a best pick-up and drop-off locations, and a sharing cost rate between rider and driver for their common path. The main idea of our approaches consists in labelling interesting nodes of a geographical map with information about drivers, in so-called buckets. Based on the information contained in these buckets, when a rider enters the system we determine a best driver, as well as a best pick-up and drop-off locations that minimize the total travel cost of rider and driver. Exact and heuristic approaches to identify a best driver, as well as a best pick-up and drop-off locations are proposed. Finally, we perform a comparative evaluation using real road network of the Lorraine region (FR) and real data provided by a local company. Experimental analysis shows a running time of a few seconds while improving participants' cost-savings and matching rate compared to the recurring ridesharing.

Keywords: Operations research · Transportation · Ridesharing · Shortest path problem · Geographical maps

1 Introduction

Transport demand management plays a very critical role in achieving greenhouse gas emission reduction targets. Particularly, the car pollution is one of the major causes of greenhouse emissions, and traffic congestion. In fact, at least half of total carbon emissions is due to private vehicles [12]. These problems negatively affect the environment, the economy, and more importantly average peoples' daily lives. Governments are conscious of the urgency to conserve the environment by investing in more environmentally friendly and safe modes of transportation like ridesharing systems.

Ridesharing can be considered as a mode of transportation in which individual travelers with similar routes and time schedules share a vehicle for a trip and

© Springer International Publishing Switzerland 2015
D. de Werra et al. (Eds.): ICORES 2015, CCIS 577, pp. 71–92, 2015.
DOI: 10.1007/978-3-319-27680-9_5

share travel costs such as gasoline cost, toll and parking fees. This also allows to reduce some of the hassles related to transportation and commuting. And at the same time, to provide measurable reductions in road congestion and energy consumption while helping to enhance the quality of our environment.

The spread of GPS-based location services using smartphone applications has led to the rapid growth of startups offering smartphone-enabled dispatch service for ridesharing vehicles. This change in communication technology has been accompanied by the development of a new concept of ridesharing, called *"dynamic or real-time ridesharing"* that consists in automatically and instantly matching riders and drivers through a network service.

In the literature, a ridesharing problem is considered in its simplest form, also called "recurring ridesharing", [1], in which a driver makes a detour to the rider's origin in order to pick him up, then drives to the rider's destination and finally goes to his own destination.

As discussed in [2], the matching opportunities of drivers and riders in the recurring ridesharing decrease when itineraries of riders and drivers diverge. Similar to [2], i.e., instead of allowing the driver to join the rider's origin to pick him up and drop him off at his destination, we allow drivers and riders to meet and to separate at intermediate locations, not necessarily at their respective destination. We also consider the *best offer selection problem*: for a given rider, we select a best driver that generates the highest cost-savings of the ride under matching constraints, and guaranteeing an efficient sharing cost rate (Fig. 1).

(a)   (b)

Fig. 1. Figure (a) represents the shortest paths of driver (green path) and the rider (red path) before matching. Figure (b) represents the new paths of the driver and the rider after the matching. Two new intermediate locations (intermediate meeting locations are pointed with two blue arrows and the common path is represented with blue path) are selected. The recurring approach of ridesharing doesn't detect this matching due to the detour time and cost of the driver (Color figure online).

2   Background

Ridesharing appears in practice under several forms depending on some characteristics. Table 1 presents some characteristics of three famous forms of ridesharing.

Table 1. Characteristics of some ridesharing systems.

Ridesharing form	Pick-up location	Drop-off location	Flexibility driver route	Driver detour constraint
Slugging	Specific stop	Specific stop	No	No
Taxi-ridesharing	Rider origin	Rider destination	Yes	No
Recurring-ridesharing	Rider origin	Rider destination	Yes	Yes

In *Slugging System*, drivers and riders line up at established locations and share rides to other established locations in order to take advantage of high occupancy vehicle lane (HOV) such as time-savings or toll savings. The driver usually positions the car so that the slugs are on the passenger side. The driver either displays a sign with the destination or simply lowers the passenger window to call out the destination. The slugs first in line for that particular destination is then picked up. Drivers maintain their original routes and don't make any detour to pick-up or drop-off riders. Generally, no money is exchanged between drivers and riders because of the mutual benefit: the car driver needs riders just as much as the slugs need a ride. Each party needs the other in order to survive. The main limitation of *slugging system* is the inflexibility of drivers' routes, which does not allow door-to-door transportation. Authors in [10] study the *slugging* problem and its generalization.

Taxi-ridesharing System denotes the use of common taxi-cabs by more than one rider serving multiple trips in the same taxi route. Contrary to taxi-cabs, the cost of a trip depends on the number of riders in the taxi. This incites riders to better use of taxi-car, which also avoids the under utilization seats during trips. Authors in [11] propose a practical taxi ridesharing service in which an organization operates a dynamic taxi ridesharing service.

Recurring Ridesharing System is similar to Taxi-ridesharing system in most characteristics. However, contrary to Taxi-ridesharing system, the driver in ridesharing is unwilling to drive anywhere in order to serve riders (limited on detour time). And usually, taxi ridesharing needs appropriate pricing mechanisms, generally more expensive than a recurring ridesharing, to incite taxi drivers. Several research has been reported recently in the fields of ridesharing, as in [1,6]. In this form of ridesharing, the pick-up and drop-off locations for the rider don't differ from his origin and destination, respectively.

The *ridesharing Problem with Intermediate Locations* can be seen as an extension of the *slugging problem* and the *recurring ridesharing problem*. Firstly, the driver's route can change to accommodate riders compared to *slugging problem*. Secondly, in contrast to the *recurring ridesharing problem*, the pick-up and drop-off locations for the rider are not necessarily his origin and destination, respectively.

The *problem of ridesharing with Intermediate Locations* is addressed in [2–4]. In [3], the authors consider the round trip ridesharing problem with an intermediate meeting location. The rider drives to the intermediate meeting location using his private car and parks it there, so in the return trip, he must be dropped off at that location to get his car back. Thus, for a given rider's request, the optimization system determines the best meeting location, the best driver in outgoing trip and the best driver in return trip passing via the intermediate meeting location where the rider's car was left. Authors develop an efficient approach where the objective is to minimize the total cost in round trip. Their approach was validated by experiments based on real data of ridesharing. The proposed approach increases the opportunity of matching between riders-drivers and then a significant reduction of the total travel cost compared to the recurring approach of round trip ridesharing. In [4], the authors consider the problem of ridesharing with intermediate locations, where the rider can use the public transportation either in order to reach the meeting location from his starting location, or to reach his ending location coming from the separate location. Their model doesn't consider the *best offer selection problem* and doesn't take into account any constraint of matchings (detour time and cost constraints).

In this paper, we consider the best offer selection problem in real-time ridesharing with intermediate locations. The objective function is to minimize the total travel cost in scenario involving transportation modes with time-independent arc costs (foot, private bike, private car ...), while ensuring that their detour costs and times remain reasonable. Thus, for a given rider's request, we determine a best driver, a best meeting and separate locations that minimize the total travel cost under constraints of detour time and detour cost of rider and driver. We suggest heuristic methods that reduce the number of shortest paths computations, based on an exact pruning algorithm.

The paper is organized as follows. Section 3 presents the problem and describes the notations used in the remaining sections. Section 4 describes the solving approaches. Section 5 proposes exact and heuristic methods for selection offer problem. Section 6 shows how to extend our approach when some parameters are fixed. Section 7 presents detailed experimental analysis of our proposed approaches. Finally, concluding remarks and future research are included in Sect. 8.

3   Problem Description and Notation

The network is represented as a directed graph $G(V, E)$, in which V represents a set of nodes and E the set of arcs. Nodes model intersections and arcs depict street segments. With each edge $(i, j) \in E$ two weights are associated. A path in a graph G is represented by a vector $\mu = (u, \ldots, v)$ of nodes in which two

Table 2. A list of notations.

Notation	Definition
$u \rightarrow v$	Shortest-path between node u and node v
$c_k(u \rightarrow v)$	Cost of a shortest-path between u and v for user k
$\tau_k(u \rightarrow v)$	Duration of a shortest-path between u and v for user k
$d(u, v)$	Distance as the crow flies between u and v
$\hat{\tau}(u \rightarrow v)$	Estimated smallest duration from u to v, such that $\hat{\tau}(u \rightarrow v) = \frac{d(u,v)}{v_{\max}}$, where $v_{\max}$ is the maximal speed
$[t_k^{\min}, t_k^{\max}]$	Departure time window of user k
Δ_k	Maximal detour time that can be accepted by user k, $\Delta_k \geq 0$, i.e. the extra-time that the driver (rider) can accept additionally to the travel time corresponding to his shortest path

successive nodes are connected by an edge of E. The cost $c(\mu)$ of path μ is the sum of costs of all edges in μ. A shortest-path between a source node u and a target node v is a path with minimal cost among all paths from u to v. In the ridesharing system, car owners (i.e. drivers) and riders specify their origins and destinations, their departure time windows, as well as their maximal detour time. Table 2 lists the notations used throughout this paper. (Some notations are introduced later.)

The starting and ending locations of drivers and riders are denoted by (s, t) and (s', t'), respectively. Furthermore, a user's request k can be either a driver's offer $k = o$, or rider's demand $k = d$. Then, an offer and a demand of ridesharing are represented by $o = (s, t, [t_o^{\min}, t_o^{\max}], \Delta_o)$ and $d = (s', t', [t_d^{\min}, t_d^{\max}], \Delta_d)$, respectively.

In [7], the authors develop a method that select offers with the smallest detours with respect to a request of ridesharing. First, they assume that network distances resemble costs. So, an offer perfectly fits a request only if origin and destination locations of driver and rider are identical. The matching in which, the driver makes a short detour to serve a rider is said to be a "reasonable fit".

Definition 1 (Reasonable fit in the Recurring Ridesharing). *An offer* $o = (s, t, [t_o^{\min}, t_o^{\max}], \Delta_o)$ *and a request* $d = (s', t', [t_d^{\min}, t_d^{\max}], \Delta_d)$ *form a reasonable fit if there exists a path* $\gamma_{od} = s \rightarrow s' \rightarrow t' \rightarrow t$ *in graph* G *with* $c_o(\gamma_{od}) \leq c_o(s \rightarrow t) + \varepsilon \cdot c_o(s' \rightarrow t')$, *with* $\varepsilon \geq 0$.

The value of ε can be seen as a sharing cost rate between the rider and the driver on the common path $s' \rightarrow t'$. So, the term $\varepsilon \cdot c_o(s' \rightarrow t')$ is the reward that rider provides to driver. In this setting, the driver is given an incentive to pick up the rider at his start location s' and to drop him off at his destination t'. Clearly, if $c_o(\gamma_{od}) - c_o(s \rightarrow t) > \varepsilon \cdot c_o(s' \rightarrow t')$, the driver does not accept the demand d. A reasonable choice of ε is a value less than 1. In other words, joining reasonable rides allows travelers to have their costs lower than those associated with traveling alone. This approach is straightforward but inflexible and unbalanced as it puts the whole detour effort on the driver. However, in order to get more opportunities of ridesharing, a rider can accept two different intermediate locations, r_1 for pick-up and r_2 for drop-off. Especially, the rider travels by his own means to the first intermediate location r_1 with a cost $c_d(s' \rightarrow r_1)$, where he will be picked up by the driver and dropped off at the second intermediate location r_2, then continue his journey from r_2 to t' on his own. Thus, the driver and the rider will share the common travel cost $c_o(r_1 \rightarrow r_2)$. If the rider rewards the driver with the amount $\varepsilon \cdot c_o(r_1 \rightarrow r_2)$, then the cost of rider is $c_d(s' \rightarrow r_1) + \varepsilon \cdot c_o(r_1 \rightarrow r_2) + c_d(r_2 \rightarrow t')$. Depending on how the value of ε is set, we distinguish two cases;

(i) **A priori approach** in which the user (driver/rider) sets the sharing cost rate ε. For instance, the rider wants to share equitably the common path cost with the driver, i.e. $\varepsilon = 0.5$.

(ii) **A posteriori approach** in which the user (driver/rider) does not set the sharing cost rate ε in advance, but the system calculates the most attractive value of ε that guarantees the best matching between the rider and the driver.

The **a priori approach** is addressed in [2]. For a given fixed value of ε, authors developed efficient heuristics that determine the best matching between the driver and the rider, while ensuring that the detour cost of the rider and the driver remains reasonable. Their objective function is to minimize the total traveling cost.

In this paper, we address the **a posteriori approach** of a ridesharing with intermediate locations, while considering additional constraints, namely, the time windows, detour time and the desired minimum cost-savings of users. More formally, we consider the general problem of ridesharing with two intermediate locations r_1 and r_2 in which driver picks up the rider at the intermediate location r_1 and drops him off at the intermediate location r_2. The rider travels on his own from s' to r_1 and from r_2 to t' (see Fig. 2).

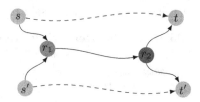

Fig. 2. The solid lines symbolize the paths of the driver and the rider in order to form a joint trip, whereas the dashed ones stand for the shortest paths of the driver and the rider.

The objective is to minimize the total travel cost, while ensuring the cost and time detour constraint for both the driver and the rider.

3.1 Matching Constraints

A matching between a driver and a rider can be established only when constraints of matching are satisfied. In our problem, we consider the timing constraint of the ride, as well as the travel cost constraint.

Timing Constraint: The timing of rides is one of the most important consideration in dynamic ridesharing. So, the choice of the new pick-up location where the driver and the rider can meet each other depends on their departure time windows. Therefore, an admissible pick-up location that forms a time synchronization between driver-rider is defined as follows:

Definition 2 (Admissible Pick-up Location). *We say that a node v is an admissible pick-up location for the driver's offer o and a rider's request d if and only if there exists $\beta \geq 0$, where*

$$\beta = \min \begin{cases} t_o^{\max} + \tau_o(s \to v) - (t_d^{\min} + \tau_d(s' \to v)) & (1a) \\ t_d^{\max} + \tau_d(s' \to v) - (t_o^{\min} + \tau_o(s \to v)) & (1b) \end{cases}$$

Equation (1a) guarantees that if the rider leaves his origin location s at time $t_d^{\min}$ to reach the intermediate pick-up location v, he must arrive no later than the latest arrival time of the driver at v. The same argument is applied to the driver in Eq. (1b). Thus, if $\beta \geq 0$, the propagation of the departure time windows of the driver and the rider at the intermediate pick-up location v will coincide. So, the earliest time that driver and rider can meet each other at this location is $\max\left\{t_d^{\min} + \tau_d(s' \to v), t_o^{\min} + \tau_o(s \to v)\right\}$.

Travel Cost Constraint: Cost saving is one of the main incentive to use a ridesharing service. However, each user (either driver or rider) may fix an extra-time to his direct duration for the use of a ride-sharing service. Thus, to guarantee these constraints, we define a reasonable fit between driver and rider as follows:

Definition 3 (Reasonable fit in Ridesharing System with Intermediate Locations). *We say that an offer $o = (s, t, [t_o^{\min}, t_o^{\max}], \Delta_o)$ and a demand $d = (s', t', [t_d^{\min}, t_d^{\max}], \Delta_d)$ form a reasonable fit if and only if there exist two intermediate locations r_1 and r_2 ($r_2 \neq r_1$) such that, r_1 is an admissible pick-up location for the offer o and demand d and*

$$c_o(s \to t) + c_d(s' \to t') - \big(c_o(s \to r_1) + c_d(s' \to r_1) + \\ c_o(r_1 \to r_2) + c_o(r_2 \to t) + c_d(r_2 \to t')\big) \geq 0 \qquad (2)$$

$$\tau_o(s \to r_1) + \tau_o(r_1 \to r_2) + \tau_o(r_2 \to t) \leq \tau_o(s \to t) + \Delta_o \qquad (3)$$
$$\tau_d(s' \to r_1) + \tau_o(r_1 \to r_2) + \tau_d(r_2 \to t') \leq \tau_d(s' \to t') + \Delta_d \qquad (4)$$

The inequality (2) represents the cost-savings of matching the offer o with the demand d. The cost-saving is defined as the difference between the travel cost of the driver and the rider when each of them travels alone without ridesharing and the total travel cost of the driver and the rider with a shared path between r_1 and r_2.

The term $(\tau_o(s \rightarrow t) + \Delta_o)$ in (3) (resp. $\tau_d(s' \rightarrow t') + \Delta_d$ in (4)) allows to limit the amount of time that the driver (resp. rider) passes in traveling. We ignore any extra-time during pick-up or drop-off of riders.

In the following lemma, we show that if the generated cost-saving is positive, then it is always possible to allocate the cost-saving among the driver and the rider so that each of them has an individual benefit.

Lemma 1. *If an offer $o = (s, t, [t_o^{\min}, t_o^{\max}], \Delta_o)$ and a demand $d = (s', t', [t_d^{\min}, t_d^{\max}], \Delta_d)$ form a reasonable fit, then there exists a sharing cost rate $\varepsilon \in [0, 1]$ such that the gains of the rider and the driver are positive.*

Proof. Assume that an offer o and a demand d form a reasonable fit. From Definition 3, there exist two intermediate locations r_1 and r_2, such that r_1 is an admissible pick-up location for the offer o and demand d and

$$c_o(s \rightarrow t) + c_d(s' \rightarrow t') - (c_o(s \rightarrow r_1) + c_d(s' \rightarrow r_1) +$$
$$c_o(r_1 \rightarrow r_2) + c_o(r_2 \rightarrow t) + c_d(r_2 \rightarrow t')) \geq 0$$

Assume that there exists a sharing cost rate $\varepsilon \in [0, 1]$ in which the common trip cost of the driver and the rider is shared according to ε, i.e., the rider rewards the driver with amount $\varepsilon \cdot c_o(r_1 \rightarrow r_2)$.

On the one hand, the driver can accept to share a ride with the rider, only if his total cost using ridesharing is less than his cost if he travels alone, i.e.,

$$c_o(s \rightarrow r_1) + (1 - \varepsilon) \cdot c_o(r_1 \rightarrow r_2) + c_o(r_2 \rightarrow t) \leq c_o(s \rightarrow t) \tag{5}$$

From (5) we obtain, $\varepsilon \geq \varepsilon_1$, where

$$\varepsilon_1 = \frac{c_o(s \rightarrow r_1) + c_o(r_1 \rightarrow r_2) + c_o(r_2 \rightarrow t) - c_o(s \rightarrow t)}{c_o(r_1 \rightarrow r_2)}$$

On the other hand, the rider accepts to be picked up and dropped off by the driver in intermediate locations, only if his total cost using ridesharing is less than his cost if he travels alone, i.e.,

$$c_d(s' \rightarrow r_1) + \varepsilon \cdot c_o(r_1 \rightarrow r_2) + c_d(r_2 \rightarrow t') \leq c_d(s' \rightarrow t') \tag{6}$$

From (6) we obtain, $\varepsilon \leq \varepsilon_2$, where

$$\varepsilon_2 = \frac{c_d(s' \rightarrow t') - c_d(s' \rightarrow r_1) - c_d(r_2 \rightarrow t')}{c_o(r_1 \rightarrow r_2)}$$

Furthermore, it is easy to see that the existence of the solution is constrained by $\varepsilon_1 \leq \varepsilon \leq \varepsilon_2$.

In order to satisfy the constraint (2), ε must be in the interval $[\varepsilon_1, \varepsilon_2]$. Thus, any value of ε in the interval $[\varepsilon_1, \varepsilon_2]$ satisfies the constraints of matching between the rider and the driver. A reasonable value of ε might be the average value of the interval $[\varepsilon_1, \varepsilon_2]$, i.e. $\varepsilon = \frac{\varepsilon_1 + \varepsilon_2}{2}$. This value corresponds to sharing cost rate between the driver and the rider on the common path.

3.2 Objective Function

As in [2], we use the term *global-path* (s, s', r_1, r_2, t, t') to describe the concatenation of paths $s \rightarrow r_1$, $s' \rightarrow r_1$, $r_1 \rightarrow r_2$, $r_2 \rightarrow t$ and $r_2 \rightarrow t'$. I.e. $(s, s', r_1, r_2, t, t') = (s \rightarrow r_1 \oplus s' \rightarrow r_1 \oplus r_1 \rightarrow r_2 \oplus r_2 \rightarrow t \oplus r_2 \rightarrow t')$. A shortest *global-path* between source nodes s, s' and target nodes t, t' is a *global-path* with minimal cost $c(s, s', r_1, r_2, t, t')$ among any *global-path* from s, s' to t, t', where

$$c(s, s', r_1, r_2, t, t') = c_o(s \rightarrow r_1) + c_d(s' \rightarrow r_1) + c_o(r_1 \rightarrow r_2)$$
$$+ c_o(r_2 \rightarrow t) + c_d(r_2 \rightarrow t') \qquad (7)$$

Thus, the objective of the ridesharing system is to determine the best intermediate locations r_1 and r_2 that minimize the shortest *global-path* such that the offer o and the demand d form a reasonable fit.

4 Solution Approach

The proposed approaches are based on computing several shortest paths. Route planning in road networks is solvable by Dijkstra's algorithm [5]. In [13], the authors note that the Dijkstra algorithm can in theory be used to find the shortest path on a road network, but for large networks it would be far too slow. Different speed-up techniques, such such as bidirectional search, goal direction, etc. In our approach, constraints defined in Sect. 3.1 will be exploited to reduce the research space of meeting locations.

4.1 Search Space of Potential Meeting Locations

The matching constraints defined in Sect. 3.1 allows us to limit the search space of potential pick-up and drop-off locations. In the following, we provide characteristics of potential pick-up and drop-off locations.

Definition 4 (Potential Pick-up and Drop-off Locations). *Let* $N^\uparrow(s)$, $N^\uparrow(s')$, $N^\downarrow(t)$ *and* $N^\downarrow(t')$ *be sets of nodes defined as follows.*

$$N^\uparrow(s) = \{v \in V \mid c_o(s \rightarrow v) \leq c_o(s \rightarrow t)$$
$$\wedge \tau_o(s \rightarrow v) + \tau_o(v \rightarrow t) \leq \tau_o(s \rightarrow t) + \Delta_o\}$$

$$N^\downarrow(t) = \{v \in V \mid c_o(v \rightarrow t) \leq c_o(s \rightarrow t)$$
$$\wedge \tau_o(s \rightarrow v) + \tau_o(v \rightarrow t) \leq \tau_o(s \rightarrow t) + \Delta_o\}$$

$$N^{\uparrow}(s') = \{v \in V \mid c_d(s' \to v) \leq c_d(s' \to t')$$
$$\wedge \, \tau_d(s' \to v) + \hat{\tau}(v \to t') \leq \tau_d(s' \to t') + \Delta_d\}$$

$$N^{\downarrow}(t') = \{v \in V \mid c_d(v \to t') \leq c_d(s' \to t')$$
$$\wedge \, \hat{\tau}(s' \to v) + \tau_d(v \to t') \leq \tau_d(s' \to t') + \Delta_d\}$$

$N^{\uparrow}(s)$ represents the set of potential pick-up locations of the driver. Indeed, if v is a pick-up location, the best situation for the driver is to share the cost $c_o(v \to t)$ with the rider as long as v satisfies the traveling time constraint $\tau_o(s \to v) + \tau_o(v \to t) \leq \tau_o(s \to t) + \Delta_o$. Thus, if a node v does not respect the cost and the traveling time constraints, then v cannot be a pick-up location. Set $N^{\downarrow}(t)$ represents the potential drop-off locations of the driver. The same reasoning is applied to $N^{\downarrow}(t)$. On the other hand, the sets $N^{\uparrow}(s')$ and $N^{\downarrow}(t')$ represent the potential pick-up and drop-off locations of the rider, respectively.

Remark that in the traveling time constraint of set $N^{\uparrow}(s')$ (resp. $N^{\downarrow}(t')$), we use $\hat{\tau}(v \to t')$ (resp. $\hat{\tau}(s' \to v)$), i.e., the traveling time corresponding to the estimated distance between two locations v and t' (resp. s' and v) using Haversine formula, instead of the shortest traveling time $\tau_d(v \to t')$ (resp. $\tau_d(s' \to v)$).

In fact, for a given node v, if $\tau_d(s' \to v) + \tau_d(v \to t') > \tau_d(s' \to t') + \Delta_d$, we can't deduce that v cannot be a pick-up location. This is due to the fact that there may be another node v', where v' is a potential drop-off location and $t_o(v \to v') + t_d(v' \to t') < \tau_d(v \to t')$.

The following lemma simultaneously characterizes the set of potential intermediate locations for both driver and rider.

Lemma 2. *A node $v \in V$ is a potential intermediate node if and only if $v \in C = C_1 \cup C_2$ where $C_1 = N^{\uparrow}(s) \cap N^{\uparrow}(s')$ and $C_2 = N^{\downarrow}(t) \cap N^{\downarrow}(t')$.*

4.2   Intermediate Meeting Locations Selection

In this section, we propose two solving methods for intermediate meeting locations selection. The first method is based on bidirectional search, and the second method is based on one-to-all shortest paths. Both methods are based on the modification of the graph G.

Once the potential intermediate locations class C has been determined, we add to the graph G two dummy nodes $S^{\star}$ and $T^{\star}$. The node $S^{\star}$ is connected to each node v, $v \in C_1$, with an arc $(S^{\star}, v)$ and the cost $c(S^{\star}, v) = c_o(s \to v) + c_d(s' \to v)$. The node $T^{\star}$ is connected to each node v, $v \in C_2$, with an arc $(v, T^{\star})$ and the cost $c(v, T^{\star}) = c_o(v \to t) + c_d(v \to t')$. The arc $(S^{\star}, v)$ represents the driver and the rider moves out from their starting locations to the pick-up node v. The arc $(v, T^{\star})$ represents the driver and the rider moves out from the drop-off node v to their ending locations. Figure 3 represents an example of transformed graph.

In the following, we describe our methods.

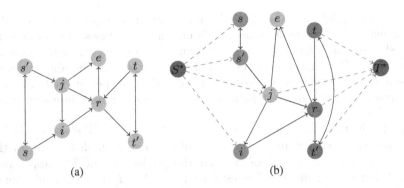

Fig. 3. Sub-figure (a) represents an example of a graph G consisting of 8 nodes. The sub-figure (b) represents the new transformed graph G' that consists in two classes C_1 (green nodes) and C_2 (blue nodes). The node e does not belong neither to C_1 nor to C_2. The node j belongs to both C_1 and C_2 (Color figure online).

Bidirectional Search Algorithm - BSA. The BSA Algorithm works with two simultaneous searches of the shortest path. One from the source node $S^\star$ and the second from the target node $T^\star$ at the same time, until "the two search frontiers meet each other". More precisely, BSA Algorithm maintains two priority queues, one for the forward search from the source $S^\star$ denoted by Q_1, as in simple Dijkstra Algorithm, and one for the backward search from the target $T^\star$ denoted by Q_2, which is a forward search in the reversed graph G^{-1}, in which each arc (u, v) of G is replaced by (v, u) in G^{-1}.

In each iteration, we settle the node with the smallest overall tentative cost, either from the source $S^\star$ or to the target $T^\star$. In order to do this, a simple comparison of the minima of the two priority queues Q_1 and Q_2 suffices. Once we settle a node v in one queue that is already settled in the other queue, we get the first tentative cost of a shortest path from $S^\star$ to $T^\star$. Its cost is the cost of the path found by the forward search from $S^\star$ to v, plus the cost of the path found by the backward search from v to $T^\star$.

Even when the two parts of the tentative solution of the forward and the backward meet each other, the concatenated solution path is not necessarily optimal. To guarantee optimality, we must continue until the tentative cost of the current minima of the queues is above the current tentative shortest path cost (which then corresponds to the cost of the shortest path). Thus, at each iteration, when the tentative cost of a shortest path from $S^\star$ to $T^\star$ is updated, we check the feasibility of constraints (2), (3) and (4), and finally we keep the admissible path that minimizes the *global-path*.

In the BSA Algorithm, when a node v is settled, then for each outgoing arc (v, u) from the node v, we check whether via this arc the node u can be reached with a cost less than or equal to the current cost of u. If yes, then the cost of u is updated to the new lower cost. This procedure is called *relaxing an arc*. The difference with the traditional *relaxing an arc* procedure of the literature is simply on updating the cost of the node even if the new cost is equal to the current cost of the node. Indeed, this modification allows us to ensure that if the shortest path from $S^\star$ to $T^\star$ contains only one intermediate node i.e., $S^\star \overset{arc}{\leadsto} r_1 \overset{arc}{\leadsto} T^\star$. Then the driver and the rider will never be able to form a reasonable fit. Specifically, the constraint (2) will never be satisfied. This case can happen when the shortest path of the driver and the rider intersect only in one location.

Finally, in the found shortest path $S^\star \to T^\star$ (i.e. $S^\star \overset{arc}{\leadsto} r_1^\star \to r_2^\star \overset{arc}{\leadsto} T^\star$), we recover the best pick-up location $r_1^\star$ and the best drop-off location $r_2^\star$ that correspond to the successor of $S^\star$ and the predecessor of $T^\star$, respectively. Figure 4 shows an example of shortest path from $S^\star$ to $T^\star$ using our relaxing arc procedure.

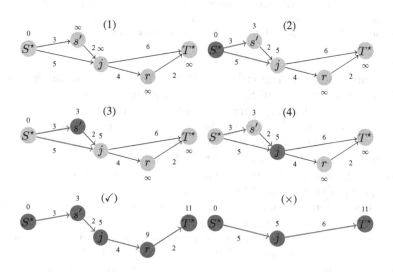

Fig. 4. Shortest path from $S^\star$ to $T^\star$ using our *relaxing arc procedure.*

Initially, all nodes except the source are labelled with tentative cost equal to infinity. Then, the tentative costs of the two neighbors s' and j are updated to 3 and 5. Secondly, once the node $S^\star$ is settled, we select the node with lower tentative cost, node s'. At this level, the path to reach the node j (neighbor of s') must be updated to $S^\star \overset{arc}{\leadsto} s' \overset{arc}{\leadsto} j$, even if the two paths have the same cost, the first one coming directly from $S^\star$, and another passing via s'. The same reasoning is applied for each node. And finally, we get the path composed of 5 nodes i.e. $S^\star \overset{arc}{\leadsto} s' \overset{arc}{\leadsto} j \overset{arc}{\leadsto} r \overset{arc}{\leadsto} T^\star$ instead of the path constituted of 3 nodes i.e. $S^\star \overset{arc}{\leadsto} j \overset{arc}{\leadsto} T^\star$ (composed of only one intermediate node).

Shortest Path One-to-All - SPOA. In this method, instead of computing one shortest path from $S^\star$ to $T^\star$, we compute for each node v in $\mathcal{C}_2$ the shortest path

$S^\star \to v$. This approach allows us to find all paths that satisfy constraint (2). Once shortest paths satisfying the constraint of *cost-savings* have been enumerated, we select the *global-path* with minimal cost that satisfies constraints (3) and (4).

The main advantage of the first method (**BSA**) lies in the fact that its running-time is less important than the second method (**SPOA**), which requires to compute all shortest paths to each node in class $\mathcal{C}_2$. On the other hand, the second method has a greater control over the constraints of *detour time* for the driver and the rider. We note that the *relaxing arc* procedure applied in this method is the same as that described above.

5   Best Offer Considerations

The approach developed above and in [2] correspond to the case where an offer and a demand are already fixed. However, in ridesharing systems, announcements of riders and drivers are arriving continuously at arbitrary times. Thus, in this section, we consider the problem of the best driver selection. Especially, given a set of drivers' offers already in the system, when a demand request d arrives in the system, the objective is to select the offer candidate which is able to fulfill the ride request.

Several selection criteria could be considered in order to find the best driver. For instance, we can choose the driver that allows the rider to arrive as soon as possible at his destination or/and that minimize their total travel time. The first objective could be conflicting with the second objective. Indeed, the driver who leads to an earliest rider arrival time at his destination is not necessary the driver that minimizes the total travel time.

To take into consideration the drivers having already began their trips, we must update their locations at each time unit. For simplicity, we consider only drivers who have not yet started their trips when the rider enters the system.

The selection rule of the best offer requires the storage of some informations for each offer entering the system: when an offer o_i is added to the system, we store all possible intermediate locations for this offer. Thus, we determine the costs $c_o(s_i \to v)$ and $c_o(v \to t_i)$. The procedure of adding offer is detailed below.

5.1   Adding an Offer

We denote by $N^\uparrow(s_i)$ and $N^\downarrow(t_i)$ the *forward search space* from a source s_i and the *backward search space* from target t_i, respectively. A forward search space $N^\uparrow(s_i)$ is a set of triplets $(v, c_{s_i}^\uparrow, \tau_{s_i}^\uparrow)$, where $(v, c_{s_i}^\uparrow, \tau_{s_i}^\uparrow)$ means that the shortest path from s_i to v has a cost $c_{s_i}^\uparrow$ and a travel time $\tau_{s_i}^\uparrow$. A backward search space $N^\downarrow(t_i)$ is a set of triplets $(v, c_{t_i}^\downarrow, \tau_{t_i}^\downarrow)$ where $(v, c_{t_i}^\downarrow, \tau_{t_i}^\downarrow)$ means that the shortest path from v to t_i has a cost $c_{t_i}^\downarrow$ and a travel time $\tau_{t_i}^\downarrow$.

Using *detour time constraint* (3) of the driver, we limit the search space. Then the triplets *(node, cost, time)* are stored in $N^\uparrow(s_i)$ and $N^\downarrow(t_i)$ without considering any demand. In order to determine the sets $N^\downarrow(t_i)$ and $N^\uparrow(s_i)$, we use a modified $A^\star$ algorithm [9]. Firstly, we determine the set $N^\downarrow(t_i)$ using reverse $A^\star$ algorithm from the destination t_i toward the origin location s_i. Then, we determine the set $N^\uparrow(s_i)$ using $A^\star$ algorithm from the origin s_i toward the destination t_i. Recall that in $A^\star$ algorithm, the nodes are retrieved from the priority queue by the sum of their tentative cost from the origin to the nodes and the value of the heuristic function from the nodes to the destination. Thus, for a given node v, the cost function used in the first $A^\star$ Algorithm to determine $N^\downarrow(t_i)$ is $\hat{\tau}(s_i \rightarrow v) + \tau_o(v \rightarrow t_i)$, whereas in the second $A^\star$ Algorithm that determines $N^\uparrow(s_i)$, we use the cost function $\tau_o(s_i \rightarrow v) + \tau_o(v \rightarrow t_i)$. Remark that in the second $A^\star$ algorithm, we use $\tau_o(v \rightarrow t_i)$ instead of the estimation duration $\hat{\tau}(v \rightarrow t_i)$, since it is already calculated in the first $A^\star$ algorithm.

We note that the traditional $A^\star$ algorithm stops after the destination has been reached. However, our $A^\star$ algorithm used to determine $N^\downarrow(t_i)$ (resp. $N^\uparrow(s_i)$) continues even when the destination s_i (resp. t_i) is reached. The algorithm stops once all nodes with cost function less than $\tau_o(s_i \rightarrow t_i) + \Delta_i$ are labelled.

Furthermore, we define the bucket $B_o(v)$ as the set which stores all drivers' trips which can pass via this location without violating the lower-bound of *detour time* constraint i.e. $\tau_o(s_i \rightarrow v) + \tau_o(v \rightarrow t_i) \leq \tau_o(s_i \rightarrow t_i) + \Delta_{o_i}$. Then, in the second algorithm $A^\star$, when a node v is labelled, we add the entries in

$$B_o(v) := B_o(v) \cup \{(i, c_{s_i}^\uparrow, \tau_{s_i}^\uparrow, c_{t_i}^\downarrow, \tau_{t_i}^\downarrow)\}. \tag{8}$$

5.2 Removing Outdated Offers

Removing offers is necessary when a ride has been matched or in case a driver wishes to retract the offer. So, to remove an offer o_i, we need to remove its entries from the buckets. In order to accelerate the time calculation of the system, we classify buckets according to calendar date. Furthermore, for each calendar date, we have not defined an order in which the entries of a bucket are stored. This makes adding operation of an offer very fast, but removing it, requires scanning the buckets. Thus, scanning all buckets is prohibitive as there are too many entries. Instead, it is faster to compute $N^\uparrow(s_i)$ and $N^\downarrow(e_i)$ in order to obtain the set of buckets which contains an entry of this offer o_i. Then, we only need to scan those buckets and remove the entries of offer o_i.

5.3 Exact Offer Selection

Let $\sigma^\star$ be the cost of the optimal global-path, initially $\sigma^\star = +\infty$. After constructing the sets $N^\uparrow(s')$ and $N^\downarrow(t')$, we scan the nodes of $N^\uparrow(s')$. For each potential intermediate location v_1 in the set $N^\uparrow(s')$, we calculate $c_o(v_1 \rightarrow v_2)$ using the forward one-to-all Dijkstra algorithm for all $v_2 \in N^\downarrow(t')$. For each offer o_i such that $(i, d_{s_i}^\uparrow, \tau_{s_i}^\uparrow, d_{t_i}^\downarrow, \tau_{t_i}^\downarrow) \in B_o(v_1)$, we scan each node v_2 in the set $N^\downarrow(t')$. Then, if i is in the set $B_o(v_2)$ such that o_i and d form a reasonable fit with v_1 (v_2) as an intermediate pick-up (drop-off) location, and the cost of the *global-path* $c(s, s', v_1, v_2, t, t')$ is less than $\sigma^\star$, then we update the optimal *global-path* cost $\sigma^\star$ to $c(s, s', v_1, v_2, t, t')$. The detail of the procedure is given in Algorithm 1.

Algorithm 1. Exact offer selection.

Require: Demand d, sets $N^\uparrow(s')$, $N^\downarrow(t')$, and $B_o(v), \forall v \in G$.

Ensure: The best driver i^*, global-path σ^*.

1: Initialization, $i^* \leftarrow -1$, $\sigma^* \leftarrow +\infty$.
2: **for all** v_1 in $N^\uparrow(s')$ **do**
3:     Using the forward one-to-all Dijkstra Algorithm, compute the cost $c_o(v_1 \rightarrow v_2)$,
       $\forall\, v_2 \in N^\downarrow(t')$.
4:     **for all** i in $B_o(v_1)$ **do**
5:         **if** v_1 is an admissible pick-up location for i and d **then**
6:             **for all** v_2 in $N^\downarrow(t')$ **do**
7:                 **if** $i \in B_o(v_2)$ **and** driver i and rider d form a reasonable fit with v_1 as
                   new pick-up location and v_2 as new drop-off location **then**
8:                     **if** $\sigma^* > c(s_i, s', v_1, v_2, t_i, t')$ **then**
9:                         $i^* \leftarrow i, \sigma^* \leftarrow c(s_i, s', v_1, v_2, t_i, t')$
10:                    **end if**
11:                **end if**
12:            **end for**
13:        **end if**
14:    **end for**
15: **end for**

Note that the runtime of Dijkstra's Algorithm using Fibonacci Heaps is bounded by $O(|V| \log |V| + |E|)$. So, the two sets $N^\uparrow(s')$, $N^\downarrow(t')$ can be computed in $O(2 \cdot (|V| \log |V| + |E|))$. Furthermore, Algorithm 1 runs in $O(|N^\uparrow(s')|(|V| \log |V| + |E|) + |N^\downarrow(t')| \cdot \sum_{v_1 \in N^\uparrow(s')} |B_o(v_1)|)$-time. Hence, the worst-case complexity of the *Exact offer selection* is $O((|N^\uparrow(s')|+2)(|V| \log |V| + |E|) + |N^\downarrow(t')| \cdot \sum_{v_1 \in N^\uparrow(s')} |B_o(v_1)|)$-time.

5.4 Heuristic Offer Selection

Even if the exact offer selection method provides a solution in polynomial time, the computation time can be too long for large instances with real road network. In particular, when the proposed method is used in real-time ridesharing system, in which a solution should be provided in a few seconds. In this section, we propose an heuristic method with a lower time complexity, that allows us to select the best offer that will be matched with the demand d.

The *heuristic offer selection* is provided by Algorithm 4 composed of Algorithms 2 and 3.

Algorithm 2 allows us to determine an offer that minimizes the *global-path*, by considering the nodes of the set $N^\uparrow(s')$ as potential pick-up locations, while the drop-off location is fixed at the rider (and driver) destination. Steps 6 and 11 of Algorithm 2 check the admissibility of the found path based on the maximum detour time allowed by the driver and the rider, as well as the generated cost-savings.

On the other hand, Algorithm 3 determines the best offer that minimizes the *global-path*, by considering the nodes of the set $N^\downarrow(t')$ as potential intermediate

---

Algorithm 2. Scan the set $N^{\uparrow}(s')$.

Require: Demand d, set $N^{\uparrow}(s')$, set $N^{\downarrow}(t')$ and $B_o(v), \forall v \in G$.

Ensure: The best driver i^*, global-path σ^*.

1: Initialization: $i^* \leftarrow -1$, $\sigma^* \leftarrow \infty$.
2: Using backward one-to-all Dijkstra Algorithm, compute $c_o(v_1 \to t')$, $\forall\, v_1 \in N^{\uparrow}(s')$.
3: **for all** v_1 in $N^{\uparrow}(s')$ **do**
4:    **for all** i in $B_o(v_1)$ **do**
5:       **if** v_1 is an admissible pick-up location for i and d **then**
6:          **if** $i \in B_o(t')$ **and** $\tau_d(s' \to v_1) + \tau_o(v_1 \to t') \leq (\tau_d(s' \to t') + \Delta_d)$ **and** $\tau_o(s_i \to v_1) + \tau_o(v_1 \to t') + \tau_o(t' \to t_i) \leq (\tau_o(s_i \to t_i) + \Delta_{o_i})$ **and** $c_o(s_i \to t_i) + c_d(s' \to t') - c(s_i, s', v_1, t', t_i, t') \geq 0$ **then**
7:            **if** $\sigma^* > c(s_i, s', v_1, t', t_i, t')$ **then**
8:               $i^* \leftarrow i$, $\sigma^* \leftarrow c(s_i, s', v_1, t', t_i, t')$
9:            **end if**
10:         **end if**
11:          **if** $\tau_d(s' \to v_1) + \tau_o(v_1 \to t_i) + \tau_d(t_i \to t') \leq (\tau_d(s' \to t') + \Delta_d)$ **and** $\tau_o(s_i \to v_1) + \tau_o(v_1 \to t_i) \leq (\tau_o(s_i \to t_i) + \Delta_{o_i})$ **and** $c_o(s_i \to t_i) + c_d(s' \to t') - c(s_i, s', v_1, t_i, t_i, t') \geq 0$ **then**
12:            **if** $\sigma^* > c(s_i, s', v_1, t_i, t_i, t')$ **then**
13:               $i^* \leftarrow i$, $\sigma^* \leftarrow c(s_i, s', v_1, t_i, t_i, t')$
14:            **end if**
15:         **end if**
16:       **end if**
17:    **end for**
18: **end for**

---

drop-off locations, while the pick-up location of the ride is fixed to the origin location of the driver (and rider). Finally, the best offer i^* is given by the best result of Algorithms 2 and 3. In order to improve the cost of the *global-path* of matching (i^*, d), we use the method described in Sect. 4.2 to find the best intermediate locations. The detail of the procedure is given in Algorithm 4.

Complexity. Algorithms 2 and 3 run in respectively, $O(|V| \log |V| + |E| + \sum_{v_1 \in N^{\uparrow}(s')} |B_o(v_1)|)$-time and $O(|V| \log |V| + |E| + \sum_{v_2 \in N^{\downarrow}(t')} |B_o(v_2)|)$-time. Then, having the sets $N^{\uparrow}(s')$, $N^{\downarrow}(t')$ and the best offer i^*, the step 4 in the Algorithm 4 can be done in $O((|V| \log |V| + |E|) + |N^{\uparrow}(s')| + |N^{\downarrow}(t')|)$-time. Finally, the *offer selection heuristic* method allows us to reduce the complexity to $O(5 \cdot (|V| \log |V| + |E|) + |N^{\uparrow}(s')| + |N^{\downarrow}(t')| + \sum_{v_1 \in N^{\uparrow}(s')} |B_o(v_1)| + \sum_{v_2 \in N^{\downarrow}(t')} |B_o(v_2)|)$.

6 Minimum Cost-Savings Rate

In some cases, either the driver or the rider may ask for a minimum rate of the saved cost in his trip. For instance, the driver requires at least 10 % of the saved cost relative to his initial travel cost. In that case, we extend model to take into account that requirement. More precisely, an offer and a demand of ridesharing will be represented by $o = (s, t, [t_o^{\min}, t_o^{\max}], \Delta_o, \sigma_o)$ and

Algorithm 3. Scan the set $N^{\downarrow}(t')$.

Require: Demand d, set $N^{\uparrow}(s')$, set $N^{\downarrow}(t')$ and $B_o(v), \forall v \in G$.

Ensure: The best driver i^*, global-path σ^*.

1: Initialization: $i^* \leftarrow -1$, $\sigma^* \leftarrow \infty$.
2: Using forward one-to-all Dijkstra algorithm, compute $c_o(s' \rightarrow v_2)$, $\forall\, v_2 \in N^{\downarrow}(t')$.
3: **for all** v_2 in $N^{\downarrow}(t')$ **do**
4:   **for all** i in $B_o(v_2)$ **do**
5:     **if** s' is an admissible pick-up location for i and d **then**
6:       **if** $i \in B_o(s')$ **and** $\tau_o(s' \rightarrow v_2) + \tau_d(v_2 \rightarrow t') \leq (\tau_d(s' \rightarrow t') + \Delta_d)$ **and** $\tau_o(s_i \rightarrow s') + \tau_o(s' \rightarrow v_2) + \tau_o(v_2 \rightarrow t_i) \leq (\tau_o(s_i \rightarrow t_i) + \Delta_{o_i})$ **and** $c_o(s_i \rightarrow t_i) + c_d(s' \rightarrow t') - c(s_i, s', s', v_2, t_i, t) \geq 0$ **then**
7:         **if** $\sigma^* > c(s_i, s', s', v_2, t_i, t')$ **then**
8:           $i^* \leftarrow i$, $\sigma^* \leftarrow c(s_i, s', s', v_2, t_i, t')$
9:         **end if**
10:        **end if**
11:      **end if**
12:      **if** s_i is an admissible pick-up location for i and d **then**
13:        **if** $\tau_d(s' \rightarrow s_i) + \tau_o(s_i \rightarrow v_2) + (\tau_d(v_2 \rightarrow t') \leq (\tau_d(s' \rightarrow t') + \Delta_d)$ **and** $\tau_o(s_i \rightarrow v_2) + \tau_o(v_2 \rightarrow t_i) \leq (\tau_o(s_i \rightarrow t_i) + \Delta_{o_i})$ **and** $c_o(s_i \rightarrow t_i) + c_d(s' \rightarrow t') - c(s_i, s', s_i, v_2, t_i, t') \geq 0$ **then**
14:          **if** $\sigma^* > c(s_i, s', s_i, v_2, t_i, t')$ **then**
15:            $i^* \leftarrow i$, $\sigma^* \leftarrow c(s_i, s', s_i, v_2, t_i, t')$
16:          **end if**
17:        **end if**
18:      **end if**
19:    **end for**
20: **end for**

Algorithm 4. Heuristic offer selection.

Require: Demand d, set $N^{\uparrow}(s')$, set $N^{\downarrow}(t')$ and $B_o(v), \forall v \in G$.

Ensure: The best driver i^*, global-path σ^*.

1: Find the best driver i_1^* in $N^{\uparrow}(s')$ using **Algorithm 2**.
2: Find the best driver i_2^* in $N^{\downarrow}(t')$ using **Algorithm 3**.
3: Select between the two drivers i_1^* and i_2^* the driver i^* with shortest global-path $\sigma^* = \min(\sigma_1^*, \sigma_2^*)$.
4: Find the best intermediate locations r_1 and r_2 between driver i^* and demand d using method described in section 4.2.

$d = (s', t', [t_d^{\min}, t_d^{\max}], \Delta_d, \sigma_d)$, respectively, where σ_o (σ_d) is the minimum percentage of cost-savings fixed by the driver (rider) relative to his shortest path. Constraint 2 of the Definition 3 is modified as follows.

$$c_o(s \rightarrow t) + c_d(s' \rightarrow t') - (c_o(s \rightarrow r_1) + c_d(s' \rightarrow r_1) + c_o(r_1 \rightarrow r_2) +$$
$$c_o(r_2 \rightarrow t) + c_d(r_2 \rightarrow t')) \geq \sigma_o \cdot c_o(s \rightarrow t) + \sigma_d \cdot c_d(s' \rightarrow t') \quad (9)$$

The values of ε_1 and ε_2 in Lemma 1 are modified in order to consider a constraint (9).

7   Computational Experiments and Discussion

In this section, we provide experimental results using the proposed methods, i.e., the two heuristics **BSA** and **SPOA**, the exact method (**EM**) and the recurring ridesharing (**CR**).

Environment. The proposed methods were implemented in C# Visual Studio 2010. The experiments were done on an Intel Xeon E5620 2.4 Ghz processor, with 8 GB RAM memory.

Offers-demands Data. In our experiments, we use real data provided by Covivo company[1]. These data concern employees of Lorraine region traveling between their homes and their work places. The real data instance is composed of 1513 participants. Among them, 756 are willing to participate in ridesharing service only as drivers. The rest of participants (i.e., 757) are willing to use other modes in order to reach the intermediate pick-up location from their origins, as well as from the intermediate drop-off location to their destinations. However, given the lack of available data of others transportation modes in the French region Lorraine, we assume that the driver and the rider have the same cost (i.e., each rider owns a private car that he might use during a part of his trip, either in order to reach the intermediate pick-up location coming from his origin, or/and reaching his destination coming from the intermediate drop-off location). Thus, the data instance is composed of 756 offers and 757 demands. The smallest and the greatest trip's distances are 2 km and 130 km, respectively.

The set of offers and demands are filtered such that we can never find a driver's offer and a rider's demand which have either the same starting or ending locations. This way will allow us to test the flexibility of our approach compared to the recurring ridesharing approach. The time window for each trip is fixed as follows. The earliest departure time and the latest departure time are fixed at 7:30 a.m. and at 8:00 a.m. respectively. The detour time of the driver (rider) is fixed to at most 20 % of his initial trip duration.

Road Network. Several geographical maps are available for geolocalised applications. The most famous ones are Google Maps, Bing Maps, Nokia Maps and OpenStreetMap[2] (OSM). Our road network of the French region Lorraine was derived from the publicly available data of OSM. It consists of 797830 nodes and 2 394 002 directed edges. Each node in the road network can be an intermediate pick-up (resp. drop-off) location in which the driver and the rider can meet (resp. separate). OsmSharp is an open-source mapping tool designed to work with OpenStreetMap-data. In our experiments, we use OsmSharp's routing and OSM data processing library to test our shortest path computation on real data set.

---

[1] http://www.covivo.fr.
[2] http://www.openstreetmap.org.

Computational Results. Two computational experiments were conducted. In the first experiment, a demand and an offer of ridesharing were fixed, and we evaluate the performance of four methods (**BSA, SPOA, EM** and **CR**) in terms of cost-savings, number of matchings and running-time (see Table 3). In this first experiment, two scenarios were tested. In the first scenario, the driver's detour time was limited to 20 %, and in the second scenario, the detour time is limited to 10 %. In each scenario, several instances were generated as follows: we scan all demands and for each demand we randomly select 10 offers, then 7570 instances were tested. For each method, the following results are reported: Gap: the deviation of algorithms with respect to the optimal solution, Time: the required CPU time in seconds, Match($\mathcal{M}$): the percentage of number matching found. In Columns *Gap* and *Time*, the average values on the 7570 instances are reported. In the row ($|\mathcal{C}|$), the average number of nodes contained in potential intermediate locations (class $\mathcal{C}$) over the 7570 instances is reported.

Table 3. Performance of our algorithm with fixed offer.

	Detour time ≤ 0.2			Detour time ≤ 0.1		
	Gap (%)	$\mathcal{M}$ (%)	Time (s)	Gap (%)	$\mathcal{M}$ (%)	Time (s)
EM	-	-	844	-	-	620
BSA	0	100	1.78	0.4	94	1.22
SPOA	0	100	2.10	0.2	97	1.58
CR	27	55.7	0.92	34	15	0.83
$|\mathcal{C}|$	742			450		

As we can see, the proposed heuristics (**BSA** and **SPOA**) detect all matchings and provide exact results for all instances when the detour time is less than 20 %. Whereas, when the detour time is less than 10 %, the Match($\mathcal{M}$) of **BSA** method decreases to 94 % and **SPOA** method decreases to 97 %. On the other hand, the **CR** approach detects at most 55.7 % when the detour time is less than 20 % and decreases more sharply to 15 % with detour time less than 10 %. The Gap of **CR** can be considered as the additional cost-savings generated by using methods with intermediate locations. The running time in **BSA** method is less than in **SPOA** method, this is due to the one-to-all Dijkstra algorithm used in **SPOA** compared to **BSA**.

In the second experiment, we evaluate the heuristic selection of offers. This allows us to know in practice, if the selected offer found using one intermediate location (either pick-up or drop-off location) is generally the same selected offer found using two intermediate locations (see Table 4). The column "Time" represents the running-time of the whole algorithm, whereas the row "Same offer (SA)" represents the success rate in which the selected offer in the step 3 (Algorithm 4) is the same offer selected by the exact method (**EM**).

From Table 4, we can see that 83 % (resp. 91 % when detour time ≤ 0.2) of selected offers found using one intermediate location are the same with selected

Table 4. Performance of the whole process.

	Detour time ≤ 0.2			Detour time ≤ 0.1		
	Gap (%)	SA (%)	Time (s)	Gap (%)	SA (%)	Time (s)
EM	-	-	9409	-	-	7803
BSA	0.7	91	3.79	2.1	83	2.64
SPOA	0.3	91	5.22	1.4	83	3.87
$\sum_{v \in C} |B(v)|$	113836			84742		

offers found using the two intermediate locations. Furthermore, when detour time ≤ 0.2, in **BSA** (resp. **SPOA**) method, the solution is found in about 3 s (resp. 5 s) compared to 9409 s for the exact method (**EM**). The Gap between the two heuristics does not exceed 2.1 % in both cases of detour time values. Our operation of adding/removing a carpool offer is compared with the approach described in [3] in terms of running-time. For this, we have fixed same offers o_1, o_2 and o_3 composed of 99, 239 and 549 buckets, respectively. These offers are added/removed from the system according to the number of offers already contained in these buckets.

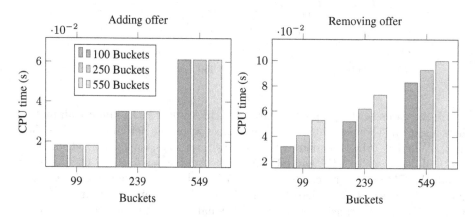

Fig. 5. Adding and removing offers depending on the number of buckets and offers in the system.

From Fig. 5, we observe that the time to add an offer o_i is independent of the number of offers already in the system, but increases with the number of buckets. The main time is spent in computing $N^{\uparrow}(s_i)$ and $N^{\downarrow}(t_i)$. However, removing an offer requires scanning these buckets. Thus, the run-time to add and remove offers remains at microseconds scale.

Figure 6 shows the gain on running time that our approach generates compared to the approach described in [3]. The observed gain is not very significant but still interesting in real-time context. The reason of positivity gain is that,

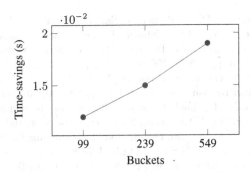

Fig. 6. Gain on running time generated by our approach compared to the approach described in [3].

the two Dijkstra algorithms used in our approach (to adding a carpool offer) are guided by heuristic functions ($A^\star$ algorithm).

8 Conclusion

The model effectively outlines the purpose and effectiveness of ridesharing and also makes it more appealing by enabling to define the meeting locations in order to share their ride. Our experiments show that our approaches enable efficient realistic ridesharing journey in large region. We used a simple objective function, which, although appropriate for certain types of users, might be redefined in several terms: the deviation from the origin-destination trip for the driver, the total travel time, etc. Our approach can integrate those features as a weighted sum in the objective function. A natural avenue for future research is accelerating our approach using Contraction Hierarchies [8].

References

1. Agatz, N., Erera, A., Savelsbergh, M., Wang, X.: Optimization for dynamic ridesharing: a review. Eur. J. Oper. Res. **223**, 295–303 (2012)
2. Aissat, K., Oulamara, A.: Dynamic ridesharing with intermediate locations. In: Symposium Series on Computational Intelligence (SSCI) (2014)
3. Aissat, K., Oulamara, A.: Round trip ridesharing with an intermediate meeting location. In: Proceedings of MOSIM 2014 (2014)
4. Bit-Monnot, A., Artigues, C., Huguet, M.J., Killijian, M.O.: Carpooling: the 2 synchronization points shortest paths problem. In: OASIcs-OpenAccess Series in Informatics, vol. 33. Schloss Dagstuhl-Leibniz-Zentrum fuer Informatik (2013)
5. Dijkstra, E.: A note on two problems in connexion with graphs. Numerische Mathematik **1**, 269–271 (1959)
6. Furuhata, M., Dessouky, M., Brunet, F.O.M., Wang, X., Koenig, S.: Ridesharing: the state-of-the-art and future directions. Transp. Res. Part B Methodol. **57**, 28–46 (2013)

7. Geisberger, R., Luxen, D., Neubauer, S., Sanders, P., Volker, L.: Fast detour computation for ride sharing. In: Erlebach, T., Lubbecke, M. (eds.) 10th Workshop on Algorithmic Approaches for Transportation Modelling, Optimization, and Systems, ATMOS 2010, pp. 88–99 (2010)
8. Geisberger, R., Sanders, P., Schultes, D., Vetter, C.: Exact routing in large road networks using contraction hierarchies. Transp. Sci. **46**, 388–404 (2012)
9. Hart, P., Nilsson, N., Raphael, B.: A formal basis for the heuristic determination of minimum cost paths. IEEE Trans. Syst. Sci. Cybern. **4**, 100–107 (1968)
10. Ma, S., Wolfson, O.: Analysis and evaluation of the slugging form of ridesharing. In: Proceedings of the 21st ACM SIGSPATIAL International Conference on Advances in Geographic Information Systems, pp. 64–73. ACM (2013)
11. Ma, S., Zheng, Y., Wolfson, O.: T-share: a large-scale dynamic taxi ridesharing service. In: IEEE 29th International Conference on Data Engineering (ICDE 2013), pp. 410–421. IEEE (2013)
12. Metz, B., Davidson, O., De Coninck, H., Loos, M., Meyer, L.: IPCC Special Report on Carbon Dioxide Capture and Storage. Cambridge University Press, Cambridge, New York (2005)
13. Sanders, P., Schultes, D.: Engineering fast route planning algorithms. In: Demetrescu, C. (ed.) WEA 2007. LNCS, vol. 4525, pp. 23–36. Springer, Heidelberg (2007)

Stochastic Semidefinite Optimization
Using Sampling Methods

Chuan Xu$^{(\boxtimes)}$, Jianqiang Cheng, and Abdel Lisser

Laboratoire de Recherche en Informatique,
Université Paris-Sud 11, 91405 Orsay Cedex, France
{chuan,cheng,lisser}@lri.fr

Abstract. This paper deals with stochastic semidefinite chance constrained problems. Semidefinite optimization generalizes linear programs,
and generally solves deterministic optimization. We propose a new sampling method to solve chance constrained semidefinite optimization problems. Numerical results are given in order to compare the performances
of our approach to the state-of-the-art.

Keywords: Linear matrix inequalities · Stochastic programming ·
Chance-constrained programming · Sample approximation · Semidefinite program

1 Introduction

It is well known that optimization models are used for decision making. In the
traditional models, all the parameters are assumed to be known, which conflicts
with many real world problems. For instance, in portfolio problems, the return
of assets are uncertain. Further, real world problems almost invariably include
some unknown parameters. Therefore, the deterministic optimization models are
inadequate and new optimization models are needed to tackle the uncertainty.
In this case, stochastic programming is the most appropriate approach to handle
the uncertainty.

As a branch of stochastic programming, chance-constrained problems (CCP)
also called probabilistic constrainted problems were first proposed in [1] to deal
with an industrial problem. The authors considered a special case of CCP where
the probabilistic constraints are used individually on each constraint. Later, [2]
generalized the model of CCP with joint probabilistic constraints and dependant
random variables. See [3–5] for a background of CCP and theoretical results
therein.

In order to circumvent CCP, we usually consider tractable approximation. For
instance, convex approximation [6,7] is a way which analytically generates deterministic convex problems which can be solved efficiently. However, it requires
to know the structure of the distribution and the structural assumptions on the
constraints. An alternative is simulation-based approach using Monte-Carlo sampling, e.g., the well-known scenario approach [8–10]. As the sampling number N

© Springer International Publishing Switzerland 2015
D. de Werra et al. (Eds.): ICORES 2015, CCIS 577, pp. 93–103, 2015.
DOI: 10.1007/978-3-319-27680-9_6

is large enough, we can ensure the feasibility of the solution. In [11], the authors developed a sampling-and-discarding approach which removes some sampling constraints from the model. They gave theoretical proofs where after discarding suitable number of constraints in the sampling model, the result remains feasible. A greedy algorithm to select the constraints to be removed was mentioned and some numerical results are shown in [12]. Recent work in [13] presented a procedure of this algorithm applied to control design.

The probabilistic problem considered is the minimum-volume invariant ellipsoid problem in control theory which can be formulated as semidefinite program with chance constrains (CCSDP). In [14], the authors proposed a convex safe tractable approximation to solve this problem. In this paper, we develop a simulation-based method [11]. For a related work on CCSDP, we refer the reader to [15–17].

The paper is organized as follows. In Sect. 2, we present mathematical formulation of the chance constrained semidefinite problem. In Sect. 3, we present simulation-based methods applied to semidefinite programs with chance constraints and introduce our method of sampling. In Sect. 4, we show numerical results on the problem in control theory. Finally, a conclusion is given in Sect. 5.

2   Chance Constrained Semidefinite Program

Conic optimization problems with chance constraints can be written as

$$(CCP) \ min\{f(x) : Pr\{F(x,\xi) \in K\} \geq 1 - \epsilon, x \in X\}$$

where $x \subseteq \mathbb{R}^n$ is a vector of decision variables, X is a deterministic feasible region, ξ is a random vector supported by a distribution $\Xi \subseteq \mathbb{R}^d$, $K \subset \mathbb{R}^l$ is a closed convex cone, $F : \mathbb{R}^n, \mathbb{R}^d \to \mathbb{R}^l$ is a random vector-valued function and ϵ is a risk parameter given by a decision maker. In this article, the probabilistic problem in our numerical tests is a bilinear semidefinite program with chance constraints. K is a positive semidefinite cone and F is a linear matrix inequality (LMI):

$$F(x,\xi) = A_0(x) + \sum_{i=1}^{m} \xi_i A_i(x) + \sum_{1 \leq j \leq k \leq m} \xi_j \xi_k B_{jk}(x)$$

where A_i, B_{jk} are symmetric matrices. Therefore, the chance constrained semidefinite program can be written as:

$$(CC_{SDP}) \ min\{f(x)_{x \in X} : Pr\{F(x,\xi) \succeq 0\} \geq 1 - \epsilon\}$$

3   Simulation-Based Approximation

3.1   Scenario Approach

The simplest method of simulation-based approximation is scenario approach. The approximation of CC_{SDP} is:

$$(CC_{SA}) \ min\{f(x)_{x \in X} : F(x,\xi^i) \succeq 0, \forall i = 1, ..., N\}$$

where N is the number of sampling, ξ^i is a random sample. CC_{SA} yields a feasible solution to CC_{SDP} with a probability of at least $1 - \beta$ for

$$N \geq \frac{2}{\epsilon} log(\frac{1}{\beta}) + 2n + \frac{2n}{\epsilon} log(\frac{2}{\epsilon}).$$

See [9] for more details.

3.2 Big-M Semidefinite Sampling Approach

In [18], the authors proposed a simulation-based method which adds a sample average constraint involving expectations of indicator functions. They showed that their simulation-based approximation method yields a feasible solution to the chance constrained problem with high confidence level. If we choose a "big-M" function with integer variables to be the indicator function, we have the following tractable approximation of CC_{SDP}:

$$(CC_{BM})\ min\ f(x)$$
$$s.t\ F(x, \xi^i) + y_i MI \succeq 0, \forall i \in 1, ..., N$$
$$\sum_{i=1}^{N} y_i \leq \epsilon \times N$$
$$x \in X, y \in \{0, 1\}^N$$

where I is an identity matrix, M is a large constant. We see that if $y_i = 1$, the constraint i is satisfied for any candidate solution x including those $x \in \{x | F(x, \xi^i) \nsucceq 0, x \in X\}$ discarded by scenario approach (CC_{SA}). This "big-M" method is less conservative than CC_{SA}, but it introduces the binary variables which increases the computation effort. The advantage of this method is that it gives a less conservative solutions.

3.3 Combination of Big-M and Constraints Discarding

In order to have a less conservative solution than the scenario approach and reduce the computational effort, our sampling method starts by solving a relaxed CC_{BM} model. We propose the continuous relaxation of the y variables in order to easily select the constraints to be removed in the sampling-and-discarding approach proposed by [11].

In our method, we suppose that the relaxed value of $y_i \in [0, 1], \forall i$ obtained by the relaxed CC_{BM} indicates the probability of discarding the constraint i. Therefore, we develop a new sampling method which combines the "big-M" approximation and sampling-and-discarding method. The main procedure consists in solving the relaxed CC_{BM} firstly and then according to the sorted value of y_i, we remove the corresponding constraints from CC_{SA} and solve the new reduced problem.

4 Numerical Experiments

We apply our method to a minimum-volume invariant ellipsoid problem in control theory [14] and compare its performance with the scenario approach and the sampling-and-discarding approach based on a greedy procedure [12].

4.1 Control System Problem

First of all, we state out the problem and its mathematical model. Suppose that we have the following discrete-time controlled dynamical system:

$$x(t+1) = Ax(t) + bu(t) \quad t = 0, 1, ...$$
$$x(0) = \bar{x}$$

where $A \in \mathbb{R}^{n \times n}$ and $b \in \mathbb{R}^n$ are system specifications, t is the index of discrete time, $\bar{x}$ is the initial state, and $u(t)$ is the control at time t. In order to keep the system stable for any A, b and possible $u(t)$, the safe region for x could be an invariant ellipsoid. An ellipsoid is expressed by:

$$E(Z) = \{x \in \mathbb{R}^n : x^T Z x \leq 1\}$$

where Z is a symmetric positive definite matrix. An invariant ellipsoid means that if $x \in E(Z)$, then $(A(x) + b) \in E(Z)$. In [19], the authors showed that the ellipsoid E(Z) is invariant if and only if there exists a $\lambda \geq 0$ such that $\begin{bmatrix} 1 - b^T Z b - \lambda & -b^T Z A \\ -A^T Z b & \lambda Z - A^T Z A \end{bmatrix} \succeq 0, \|A\| < 1$. For this problem, we prefer to have a smaller safe region for x to ensure the stability. Thus, this control problem is equivalent to finding the minimum-volume of an invariant ellipsoid which could be formulated as a bilinear semidefinite programming problem. If we consider the chance constrained case, the model should be: $\{CC_{MVIE}(\lambda), \lambda \in D\}$[14].

$$CC_{MVIE}(\lambda):$$

$$max \; w$$

$$s.t \; w \leq (det Z)^{1/n})$$

$$Pr\left\{ \begin{bmatrix} 1 - b^T Z b - \lambda & -b^T Z A \\ -A^T Z b & \lambda Z - A^T Z A \end{bmatrix} \succeq 0 \right\} \geq 1 - \epsilon$$

$$Z \succeq 0$$

where $D = \{0.00, 0.01, ..., 0.99, 1.00\}$ is a finite set. We assume that the system could be disturbed by some random noise. Like the design of numerical experiment in [14], b is corrupted and $b_i = \bar{b}_i + \rho \xi_i, \forall i = 1, ..., N$ where $\bar{b} \in \mathbb{R}^N$ is the nominal value, $\rho \geq 0$ is a fixed parameter to control the level of perturbation, ξ_i is a standard Gaussian random variable of sample i.

4.2 Sampling Procedure

4.2.1 Scenario Approach

We generate N random samples and solve the following model $\{CCSC(\lambda),$ $\lambda \in D\}$:

$$CCSC(\lambda):$$

$$max\ w$$

$$s.t\ w \leq (detZ)^{1/n})$$

$$\begin{bmatrix} 1 - b_i^T Z b_i - \lambda & -b_i^T Z A \\ -A^T Z b_i & \lambda Z - A^T Z A \end{bmatrix} \succeq 0,$$

$$\forall i = \{1, ..., N\}$$

$$Z \succeq 0$$

4.2.2 Greedy Procedure for Sampling-and-Discarding Method

For each $\lambda \in D$, we apply a greedy and randomized constraint removal procedure [12] to the sample counterpart (SP) of $CC_{MVIE}(\lambda)$ [11].

$$CCSP(\lambda):$$

$$max\ w$$

$$s.t\ w \leq (detZ)^{1/n})$$

$$\begin{bmatrix} 1 - b_i^T Z b_i - \lambda & -b_i^T Z A \\ -A^T Z b_i & \lambda Z - A^T Z A \end{bmatrix} \succeq 0,$$

$$\forall i = \{1, ..., N\} - \mathcal{A}$$

$$Z \succeq 0$$

where $\mathcal{A}$ is the set of the indexes of the k removed constraints.

The greedy removal procedure iteratively removes k constraints. At each iteration i, we solve a $CCSP(\lambda)$ with $\mathcal{A}_{i-1}$ to determine the set of n_i active constraints. Then we randomly choose one of these active constraints such that constraint c has $\mathcal{A}_i = \mathcal{A}_{i-1} \cup \{c\}$ for the following iteration $i + 1$.

4.2.3 Big-M Procedure for Sampling and Discarding Method

Our sampling method is twofold. First, we solve a relaxed "big-M" model $CCRBM(\lambda)$ and obtain the solution of the relaxed binary variable y:
$CCRBM(\lambda)$

$$max\ w$$

$$s.t\ w \leq (detZ)^{1/n})$$

$$\begin{bmatrix} 1 - b_i^T Z b_i - \lambda & -b_i^T Z A \\ -A^T Z b_i & \lambda Z - A^T Z A \end{bmatrix} + y_i M I \succeq 0,$$

$$\forall i = 1, ..., N$$

$$\sum_{i=1}^{N} y_i \leq \epsilon \times N$$

$$Z \succeq 0$$

$$0 \leq y_i \leq 1, \forall i = 1, ..., N$$

We sort the elements of y in decreasing order and take the first k indexes into set $\mathcal{A} = \{i_1, ..., i_k\}$. Then, we solve $CCSP(\lambda)$ using the following algorithm:

Algorithm.

---

1. For each $\lambda \in D$:
   (a) Solve $CCRBM(\lambda)$ and get the relaxed values of y,
   (b) Build the set $\mathcal{A}$ of removed constraints according to y,
   (c) Solve $CCSP(\lambda)$, and let $v(\lambda)$ be the objective value and $Z(\lambda)$ be the corresponding solution.
2. Return $Z(\lambda^*)$ as the optimal solution, where $\lambda^* = argmax_{\lambda \in D} v(\lambda)$.

4.3 Numerical Experiments

4.3.1 Data
We use the same instances as [14]. We have two group of data.

$$Data\,1 : A = \begin{bmatrix} -0.8147 & -0.4163 \\ 0.8167 & -0.1853 \end{bmatrix}, \bar{b} = \begin{bmatrix} 1 \\ 0.7071 \end{bmatrix},$$

$$\epsilon = 0.05, \rho = 0.01, \beta = 0.05$$

$$Data\,2 : A = \begin{bmatrix} 0 & 2 & 0 & 0 & 0 \\ 0 & 0 & 0.0028 & 0.0142 & 0 \\ 0 & 0 & 0 & 1 & 0 \\ 0 & 0 & -0.0825 & -0.4126 & 0 \\ 1 & 0 & 0 & 0 & 0 \end{bmatrix},$$

$$\bar{b} = \begin{bmatrix} 0 \\ 0.0076 \\ 0 \\ -0.1676 \\ 0 \end{bmatrix}, \epsilon = 0.03, \rho = 0.001, \beta = 0.05$$

where β is a confidence parameter which is needed to decide the sample size N and the number of removed constraints k.

4.3.2 Selecting the Sample Size and the Number of Constraints to Be Removed

For data 1, we consider four sample sizes N ranging from 400 up to 1000. The number of constraints to be removed is calculated as following:

$$k = \lfloor \epsilon N - d + 1 - \sqrt{2\epsilon In \frac{(\epsilon N)^{d-1}}{\beta}} \rfloor,$$

where d is the dimension of variable Z. It has been proven in [11] that with this value of k, the solution obtained by $CCSP(\lambda)$ (with optimal removal) is feasible to $CC_{MVIE}(\lambda)$ with high probability $1 - \beta$. As choosing the optimal set of constraints to be removed is an NP-hard problem, the solution that we obtain with our procedure can not ensure conservativeness. Therefore, we vary the ratio of k/N from 0.03 to 0.05 to study the influence of k on the result.

For data 2, we consider three sample sizes N ranging from 1000 to 1400 with k calculated as in (Sect. 4.3.1). In addition, we set the ratio of k/N to be 0.02 and 0.03 for each sample size respectively.

4.4 Numerical Results

All the experiments are carried out using MATLAB R2012b on a Windows 7 operating system with i7 CPU @ 2 GHz and 4 GB of RAM. The computations are performed using CVX 2.1 with semidefinite program solver $SeDuMi$.

Tables 1 and 2 provide the computational results of Data 1 and Data 2 respectively. N presents the sampling number. k is the number of removal constraints and k/N is the corresponding ratio. We use the average linear size measure, which is defined as $ALS(E(Z)) = (Vol_n(E(Z))^{1/n}$ [14], to evaluate the volume of the ellipsoid. The smaller the volume of the ellipsoid is, the smaller is the average linear size of the ellipsoid. The columns SC, $Greedy$, $BMSP$ give the average linear size of the ellipsoid obtained by the scenario approach (Sect. 4.2.1), the greedy approach (Sect. 4.2.2) and our method (Sect. 4.2.3) respectively. $(1 - Vio)$ shows the satisfaction rate of each solution estimated under 100000 simulated random samples. Gap presents the gap between the solution of the current method and the solution of the scenario approach. Table 3 shows the CPU time expressed in seconds. The columns SC, $Greedy$, $BMSP$ show the average CPU time of all tests in Tables 1 and 2 when applying the scenario approach, the greedy approach and our method respectively.

We observe that the real violation is significantly below 5 % and 3 % respectively in Tables 1 and 2. It is easy to see that as k increases, we obtain a better solution both with greedy method and with our $BMSP$ method; and the violation of the solution is larger. The reason is more we remove the constraints, larger is the feasible set of $CC_{SP}(\lambda)$ and more elements of $CC_{SC}(\lambda)$ are violated.

In Table 1, for each sampling value N, $BMSP$ obtains better solution than $Greedy$ with smaller final value (average linear size of the ellipsoid) and a larger violation which is below 5 %. For the greedy method, the gap is between 0.5 ‰– 10.4 ‰, compared with scenario approach. While for our method, the gap is between 2.7‰–11.8‰. Figure 1 gives an illustration of the final value obtained by *Greedy* and *BMSP* for different values of k for 400 samples. In Fig. 2, we compare the violation of *Greedy* and *BMSP*. We observe that the increasing rate of violation is nearly the same. Figure 3 shows the local view of ellipsoid for Data 1 obtained by the scenario approach, the greedy approach and our method with $N = 400$ and $k = 20$. We can see that the ellipsoid obtained by our method has the smallest volume.

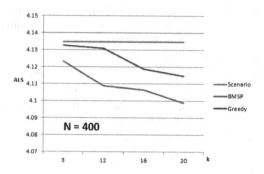

Fig. 1. Comparison of average linear size for Data 1.

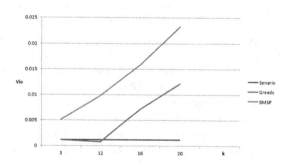

Fig. 2. Comparison of violation ratio for Data 1.

In Table 2, we obtain a *Gap* more significant than the previous one on Data 1. For the case where k is chosen by (Sect. 4.3.1), our method obtains a better gap than the *Greedy* method from 0.2 % up to 0.6 % improvement. While for other values of k, their related *gaps* are very similar.

The advantage of our method compared with *Greedy* procedure is on the computing time. In the *Greedy* procedure, we need to solve $k + 1$ times the semidefinite program $CC_{SP}(\lambda)$ in order to decide removal constrains, while in our method, we only need to solve 2 semidefinite programs. Therefore, we observe from Table 3 that *BMSP* uses significantly less CPU time than *Greedy* and

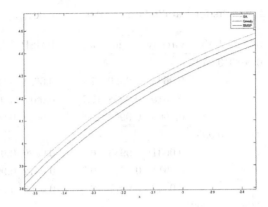

Fig. 3. Local view of chance-constrained invariant ellipsoid of Data 1 with $N = 400$, $k = 20$.

almost twice CPU time than the scenario approach. However, as a counterpart of the CPU time, we obtain better solution than scenario approach.

Table 1. Results for Data 1 with $\epsilon = 0.05, \rho = 0.01, \beta = 0.05$.

N	k	k/N	SC	1-Vio	Greedy	1-Vio	Gap(‰)	BMSP	1-Vio	Gap(‰)
400	-		4.1348	0.9988	-	-	-	-	-	-
	3	0.008			4.1328	0.9988	0.5	4.1234	0.9948	2.7
	12	0.030			4.1309	0.9992	0.9	4.1090	0.9902	6.3
	16	0.040			4.1190	0.9928	3.8	4.1065	0.9842	6.9
	20	0.050			4.1148	0.9818	4.8	4.0988	0.9767	8.8
600	-		4.1438	0.9988	-	-	-	-	-	-
	9	0.015			4.1098	0.9884	8.2	4.1095	0.9892	8.3
	18	0.030			4.1060	0.9829	9.1	4.1025	0.9811	10.0
	24	0.040			4.1050	0.9835	9.3	4.0976	0.9744	11.2
	30	0.050			4.1043	0.9799	9.5	4.0962	0.9720	11.5
800	-		4.1482	0.9998	-	-	-	-	-	-
	15	0.019			4.1151	0.9891	7.9	4.1138	0.9923	8.2
	24	0.030			4.1106	0.9917	9.0	4.1066	0.9859	10.0
	32	0.040			4.1083	0.9883	9.6	4.1028	0.9781	10.9
	40	0.050			4.1047	0.9846	10.4	4.0990	0.9776	11.8
1000	-		4.1455	0.9994	-	-	-	-	-	-
	22	0.022			4.1228	0.9968	5.4	4.1124	0.9889	7.9
	30	0.030			4.1221	0.9938	5.6	4.1066	0.9865	9.4
	40	0.040			4.1144	0.9916	7.5	4.1027	0.9791	10.3
	50	0.050			4.1050	0.9861	9.7	4.0974	0.9734	11.6

Table 2. Results for Data 2 with $\epsilon = 0.03, \rho = 0.001, \beta = 0.05$.

N	k	k/N	SC	1-Vio	Greedy	1-Vio	Gap(%)	BMSP	1-Vio	Gap(%)
1000	-		0.0689	0.9995	-	-	-	-	-	-
	14	0.014			0.0634	0.9980	7.9	0.0631	0.9966	8.5
	20	0.020			0.0615	0.9958	10.7	0.0613	0.9920	11.1
	30	0.030			0.0603	0.9908	12.5	0.0603	0.9915	12.4
1200	-		0.0677	0.9994	-	-	-	-	-	-
	6	0.013			0.0631	0.9933	6.7	0.0629	0.9970	7.1
	24	0.020			0.0611	0.9925	9.7	0.0612	0.9917	9.6
	36	0.030			0.0592	0.9877	12.5	0.0596	0.9890	12.0
1400	-		0.0664	0.9992	-	-	-	-	-	-
	17	0.012			0.0617	0.9958	7.1	0.0615	0.9943	7.3
	28	0.020			0.0603	0.9943	9.2	0.0605	0.9933	8.9
	42	0.030			0.0592	0.9868	10.8	0.0596	0.9927	10.3

Table 3. Average CPU time.

		Data 1			Data 2	
	SC	Greedy	BMSP	SC	Greedy	BMSP
CPU time	13.6	201.5	23.3	251.7	4955.2	521.4

5   Conclusion

In this paper, we introduce a new simulation-based method to solve stochastic chance constrained program. This method is a combination of Big-M relaxation and a sampling-and-discarding method. We apply this method to semidefinite programming problem in control theory. The numerical results show that our method provides better solutions within a reasonable CPU time.

References

1. Charnes, A., Cooper, W.W., Symonds, G.H.: Cost horizons and certainty equivalents: an approach to stochastic programming of heating oil. Manag. Sci. **4**, 235–263 (1958)
2. Prékopa, A.: On probabilistic constrained programming. In: Proceedings of the Princeton symposium on mathematical programming, pp. 113–138. Princeton University Press, Princeton (1970)
3. Dentcheva, D., Prékopa, A., Ruszczynski, A.: Concavity and efficient points of discrete distributions in probabilistic programming. Math. Program. **89**, 55–77 (2000)
4. Prékopa, A.: Probabilistic programming. In: Handbooks in Operations Research and Management Science, vol. 10, pp. 267–351 (2003)

5. Henrion, R., Strugarek, C.: Convexity of chance constraints with independent random variables. Comput. Optim. Appl. **41**, 263–276 (2008)
6. Nemirovski, A., Shapiro, A.: Convex approximations of chance constrained programs. SIAM J. Optim. **17**, 969–996 (2006)
7. Nemirovski, A.: On safe tractable approximations of chance constraints. Eur. J. Oper. Res. **219**, 707–718 (2012)
8. Calafiore, G., Campi, M.C.: Uncertain convex programs: randomized solutions and confidence levels. Math. Program. **102**, 25–46 (2005)
9. Calafiore, G.C., Campi, M.C.: The scenario approach to robust control design. IEEE Trans. Autom. Control **51**, 742–753 (2006)
10. Nemirovski, A., Shapiro, A.: Scenario approximations of chance constraints. In: Calafiore, G., Dabbene, F. (eds.) Probabilistic and Randomized Methods for Design Under Uncertainty, pp. 3–47. Springer, London (2006)
11. Campi, M.C., Garatti, S.: A sampling-and-discarding approach to chance-constrained optimization: feasibility and optimality. J. Optim. Theory Appl. **148**, 257–280 (2011)
12. Pagnoncelli, B.K., Reich, D., Campi, M.C.: Risk-return trade-off with the scenario approach in practice: a case study in portfolio selection. J. Optim. Theory Appl. **155**, 707–722 (2012)
13. Garatti, S., Campi, M.C.: Modulating robustness in control design: principles and algorithms. IEEE Control Syst. **33**, 36–51 (2013)
14. Cheung, S.S., Man-Cho So, A., Wang, K.: Linear matrix inequalities with stochastically dependent perturbations and applications to chance-constrained semidefinite optimization. SIAM J. Optim. **22**, 1394–1430 (2012)
15. Ariyawansa, K., Zhu, Y.: Chance-Constrained Semidefinite Programming. Technical report, DTIC Document (2000)
16. Yao, D.D., Zhang, S., Zhou, X.Y.: LQ Control without Riccati Equations: Stochastic Systems. FEW-Econometrie en besliskunde (1999)
17. Zhu, Y.: Semidefinite Programming Under Uncertainty. Ph.D. thesis, Washington State University (2006)
18. Luedtke, J., Ahmed, S.: A sample approximation approach for optimization with probabilistic constraints. SIAM J. Optim. **19**, 674–699 (2008)
19. Nemirovski, A.: Lectures on modern convex optimization. In: Society for Industrial and Applied Mathematics (SIAM), Citeseer (2001)

Evaluation of Partner Companies Based on Fuzzy Inference System for Establishing Virtual Enterprise Consortium

Shahrzad Nikghadam[1]([✉]), Bahram LotfiSadigh[1], Ahmet Murat Ozbayoglu[2], Hakki Ozgur Unver[3], and Sadik Engin Kilic[4]

[1] Mechanical Engineering Department, Middle East Technical University, Ankara, Turkey
shahrzad.nikghadam@metu.edu.tr, Bahram.LotfiSadigh@gmail.com
[2] Computer Engineering Department,
TOBB University of Economics and Technology, Ankara, Turkey
{mozbayoglu,hounver}@etu.edu.tr
[3] Mechanical Engineering Department,
TOBB University of Economics and Technology, Ankara, Turkey
[4] Mechanical Engineering, Atilim University, Ankara, Turkey
engin.kilic@atilim.edu.tr

Abstract. Virtual Enterprise (VE) is one of the growing trends in agile manufacturing concepts. Under this platform companies with different skills and core competences are cooperate with each other in order to accomplish a manufacturing goal. Success of VE, as a consortium, highly depends on the success of its partners. So it is very important to choose the most appropriate companies to enroll in VE. In this study a Fuzzy Inference System (FIS) based approach is developed to evaluate and select the potential enterprises. The evaluation is conducted based on four main criteria; unit price, delivery time, quality and past performance. These criteria are considered as inputs of FIS and specific membership functions are designed for each. By applying fuzzy rules the output of the model, partnership chance, is calculated. In the end, the trustworthy of the model is tested and verified by comparing it with fuzzy-TOPSIS technique providing a sample.

Keywords: Virtual enterprise · Partner selection · Fuzzy Inference System

1 Introduction

In todays increasingly competitive dynamic global market, traditional manufacturing concepts cannot satisfy the diverse customer demands. As Small and Medium Sized Enterprises (SMEs) are highly specialized in few manufacturing fields they are unable to produce high value added final products [4]. Virtual Enterprise (VE) provides a synergistic virtual cooperation platform for SMEs

© Springer International Publishing Switzerland 2015
D. de Werra et al. (Eds.): ICORES 2015, CCIS 577, pp. 104–115, 2015.
DOI: 10.1007/978-3-319-27680-9_7

to share their resources and core competencies without losing the flexibility and information security. VE is a network based temporary alliance between independent and geographically dispersed multiple enterprises to share skills, diverse capabilities and resources, in order to catch business opportunities [10]. This kind of consortium will help companies for responding quickly to unanticipated demands from customers.

Lifecycle of VE consists of three main phases; Formation, Operation and Dissolution. The Formation phase of a VE is usually triggered by a request for quote from customer. Based on the project, design specifications and manufacturing requirements, production processes are decomposed into individual task(s). The main step of VE formation phase is to select the best partners to participate in forthcoming VE consortium. In order to fulfil the project, each individual task should be completed by selected VE partner(s). After the operation phase of VE which includes manufacturing and assembly processes, getting customers consent and achieving the goal(s) VE project is finalized and it can be dissolved. The most important part of VE formation phase is the partner selection step. In order to form up a successful VE consortium it is crucial to select the most appropriate partners from list of potential partners registered in virtual breeding environment. This is why there are lots of researches conducted in this field. However, due to neglecting the dynamic nature of VE and heterogeneity of customer preferences (decision making criteria), much of the proposed methods are not generic solutions and cannot be implemented directly in different decision making problems.

Partner selection is not a simple optimization problems [7]. Regarding the fact that, it is very difficult to express the qualitative criteria with precise values in digits and considering the nature of quantitative criteria which are represented in numbers, handling the quantitative criteria mathematically is much easier than including qualitative criteria in mathematical models [9]. The other difficulty of decision making is that it involves conflicting criteria. If there is a potential partner with best score in all criteria surely that company is the best; however generally this is not the case in practical applications. For instance a high quality product usually comes with expensive price. Hence there is an inevitable trade-off between criteria which is done on the basis of customers preferences.

Importance of partner selection problem along with complexity of this subject drew the attention of many researchers. Some approaches use Artificial Intelligence techniques such as Genetic Algorithm to solve the partner selections mathematical model [3], where Sari et al. propose Analytic Hierarchy Process (AHP) to perform pairwise comparisons between criteria and alternatives [7]. In these methodologies quantitative criteria are assigned with a crisp value, neglecting the subjective nature of them. In contrast, most of the papers in the literature are hybrid fuzzy approaches which are capable of handling the imprecision of input data. Mikhailov and Fei propose Fuzzy-AHP and Fuzzy-TOPSIS methods respectively [6,9].

In a study conducted by Bevilacqua and Petroni fuzzy logic is employed in specifying the relative importance (weight) given to criteria and in determining the impact of each supplier on the attributes considered [1]. Yet this study is conducted in the field of supplier selection of supply chain management (SC) and there is insufficient research for applying fuzzy logic approach in partner selection problem of VE.

Selection of partner enterprises in creation of virtual enterprise has much in common with supplier selection of supply chain management. They both evaluate the companies and try to find the best alternative with respect to number of factors. However they are not completely identical. VE is more dynamic in comparison to SC. Supplier selection of SC designed for a specific set of processes, while VE can emerge for fulfilling different types of projects and customers so VE is more dynamic in comparison to SC.

The method proposed in this paper is based on applying FIS to deal with uncertainty of the problem; in addition it considers criterion-specific membership functions which is a fact neglected in the literature to the best of our knowledge. The remainder of this paper is organized as follows: Sect. 2 reviews some background information about FIS. Section 3 explains and discusses the developed model in details. An illustrative example is presented in Sect. 4 and the results of proposed model is compared with fuzzy-TOPSIS model. Conclusions are discussed and future research scopes are recommended in the last section.

2   Fuzzy Inference System

Lotfi A. Zadeh published the theory of fuzzy set mathematics in 1965 and fuzzy logic by extension [11]. Fuzzy set is a valid supporting tool to overcome uncertainty [1].

FIS is a popular reasoning framework based on the concepts of fuzzy set theory, fuzzy logic and fuzzy IF-THEN rules. FISs make decisions based on inputs in the form of linguistic variables derived from membership functions. These variables are then matched with the preconditions of linguistic IF-THEN rules called fuzzy rules, and the response of each rule is obtained through fuzzy implication as a crisp value [8].

Mamdani fuzzy inference is the most commonly used inference method introduces by Mamdani in 1975 [5]. The FIS involves four steps: 1. Fuzzification of input variables 2. Rule Evaluation 3. Aggregation of the rule outputs 4. Defuzzification.

The first step of FIS is calculating the membership degree of inputs to their belonging fuzzy sets. In the second step fuzzified values of inputs are used to evaluate fuzzy rules. Fuzzy rules are contain fuzzy operators (AND or OR). The next step is aggregating the fuzzy outputs of all rules. The last step of FIS is defuzzifying the output, conclude the final crisp value and rank the results.

3   Partner Selection Model

Among tens of aspects to evaluate the partners to join virtual enterprise, in this research, four main criteria are taken into account; proposed unit price, proposed delivery time, companys products quality and its past performance. According to the industry experts, these four criteria are believed to be the most essential aspects to evaluate the enterprises.

First two criteria are proposed by each enterprise during negotiation process. The values of last two criteria are imported from quality and performance evaluation models which are available in the system. The proposed values of price and delivery time are normalized using Euclidean normalization method. After calculating the values of enterprises for each criterion, these values are fuzzified with respect to the corresponding membership functions. Due to different nature of each criterion, different sets and membership functions are defined.

3.1 Input Variables of Fuzzy Inference System

First step to implement the model is to translate the linguistic variables into fuzzy numbers through defining the appropriate membership functions. Using different membership functions to calculate the value of each criterion is the novelty introduced in this paper, compared to the researches in the literature. i.e. different types of membership functions are selected in order to fit the actual pattern of each factor.

Unit Price. Since the price proposals are absolute values, the corresponding membership functions must be linear in order to maintain the competitiveness between candidates. Even a dollar less, means cheaper price. This should not be ignored in fuzzification process. So, three triangular membership function are used to model the fuzzy behaviour of unit price proposed by enterprises, as shown in Fig. 1. The membership function are as follows:

Inexpensive $(0; 0; 0.5)$
Moderate $(a_1; 0.5; b_1)$
Expensive $(0.5; 1; 1)$

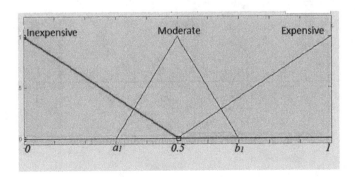

Fig. 1. Unit prices membership functions.

Delivery Time. The membership functions of delivery time are also linear. However, they consist of domains which, within that range, the fuzzified values of scores are equal. Generally Project Evaluation Review Technique (PERT) is used to calculate the Earliest Finish and Latest Finish. The range between these two due dates is a favorable domain.

If a task cannot be completed on time it will be back order charging some penalties [12]. If lateness exceeds, the order will be lost and it cannot be compensated, this is a domain which membership function has a constant value equal to one. As too early delivery imposes storage costs its trend is similar to late delivery. Trapezoidal membership function is providing all the characteristics required to model the delivery time. The membership functions are shown in Fig. 2.

Too Early $(0; 0; a_2; c_2)$
Favorable $(b_2; d_2; e_2; g_2)$
Late $(f_2; h_2; 1; 1)$

Quality. Similar to delivery time, there are constant valued domains for membership functions of quality. Quality specifications are generally defined by an acceptable "range", all the values within these limits are satisfactory. Furthermore, considering the marginality of human decisions bell-shaped membership function are most suitable to model the problem as shown in Fig. 3. Parameters of a_3 and b_3 are determining the shape of the curves.

Less than Required $(a_3; b_3; 1.5; 0)$
Satisfactory $(a_3; b_3; 0.5)$
More than Required $(a_3; b_3; 1)$

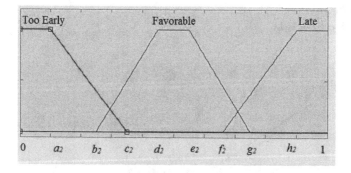

Fig. 2. Delivery times membership functions.

Past Performance. Past performance of a company is a criterion representing the customers satisfaction degree, which is influenced by several factors such as

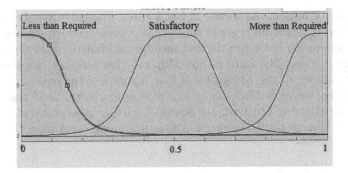

Fig. 3. Qualitys membership functions.

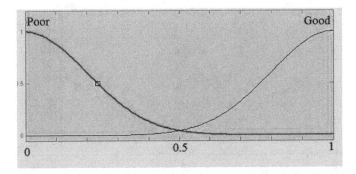

Fig. 4. Past performance's membership functions.

after sale service; respond to changes; communication openness and etc. Similar to the nonlinear trend of quality, past performances membership function is not linear due to marginality. Increasing customers consent is more demanding at higher scores. So two simple Gaussian membership functions are defined. These membership functions are shown in Fig. 4. Parameter a_4 specifies the shape of the curves

Poor $(a_4; 0)$
Good $(a_4; 1)$

3.2 Output Variable of Fuzzy Inference System

Output variable of this model is partners acceptance chance to join VE consortium. This model reveals a method to calculate this value by getting two bidding proposals (price and delivery time) and two performance evaluation value from enterprise background (quality and enterprise past performance). An enterprise with competitive proposals and good background will have higher chance to be picked out as a winner to take role in virtual enterprise rather than other rival enterprises.

Partnership Chance. As shown in Fig. 5 three triangular membership functions are used to define fuzzy set of the output. Usually enterprises violating the project requirements belong to the first membership function and their partnership chance are low. The third membership function members are those which can satisfy almost all the necessities of four inputs and the enterprises belonging to this set are most likely to be accepted as partner. While the members of second membership function, are potential partner enterprises which cannot classified in first or third membership function groups and have the medium partnership chance.

Low $(0; 0; 0.5)$
Medium $(a_5; 0.5; b_5)$
High $(0.5; 1; 1)$

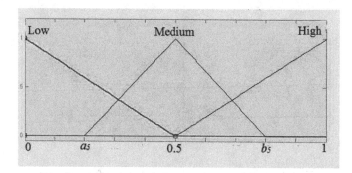

Fig. 5. Partnership chance's membership functions.

3.3   Fuzzy Rules

Once the inputs are fuzzified, fuzzy rules should be defined. Fuzzy rules are made up of linguistic statements which describe how to make decisions considering the inputs.

If (input 1 is membership function1) AND/OR (input 2 is membership function 2) THEN (Output n is membership function n)

Rules are established based on customer preferences. By asking the decision maker(s) to fill the questionnaire, the relative importance of each criterion is extracted. For instance an enterprise producing the low quality product which do not met the system predefined specifications or unable to get customer consent about delivery time is not competitive, has low partnership chance.

All the possible combinations should be considered for constructing fuzzy rules to ensure the validity of the model. More descriptive fuzzy rules will be presented for the case of our study. Establishing the IF-THEN rules are the most important step of the method since even a single improper rule will cause untrustworthy results.

According to these fuzzy rules, fuzzy inputs will be combined and evaluated by Mamdanis FIS to find the partnership chance as this models output

4   Illustrative Example

In order to illustrate the application of FIS based partner evaluation methodology to form up a VE consortium a simplified example is presented. Assume VE has received an order from a customer to manufacture the component shown in Fig. 6.

Fig. 6. Component sketch for illustrative example.

There are five candidate enterprises eager to take role in project. Bidding starts in order to identify the best proposal from the best company. Bidding proposals and candidates scores are shown in Table 1. The values of input variables, price and delivery time are proposed by each company, while the scores of quality and past performance which are called from system's data base are out of 10. Table 2 shows the normalized values of Table 1.

Table 1. Scores of enterprises with respect to criteria.

Company name	Unit price ($)	Delivery time (days)	Quality (out of 10)	Past performance (out of 10)
Company A	1600	10	5	7
Company B	1800	9	5	5
Company C	2000	12	6	6
Company D	1800	11	6	6
Company E	1500	10	7	8

The evaluation procedure will be based on FIS approach presented in previous section. The normalized values are fuzzified according to the membership

Table 2. Normalized scores of enterprises with respect to criteria.

Company name	Unit price	Delivery time	Quality	Past performance
Company A	0.409	0.428	0.382	0.483
Company B	0.460	0.385	0.382	0.345
Company C	0.511	0.514	0.459	0.414
Company D	0.460	0.471	0.459	0.414
Company E	0.384	0.428	0.535	0.552

functions specified for each criterion. Regarding the customer preference based-fuzzy rules shown in Table 3, companies partnership chance is evaluated and tabulated in Table 4.

With three levels for unit price, delivery time, quality and two levels for past performance there are total of $(3 \times 3 \times 3 \times 2) = 54$ possible combinations if all the rules are defined by AND operator. Though, to simplify the rule list, one rule is defined as below;

IF (Delivery time is Late) OR (Quality is Less than Required) THEN (Partnership Chance is Low)

By doing this, just two levels are left for delivery time and quality. As $(3 \times 2 \times 2 \times 2) + 1 = 25$, all the possible combinations are included by defining 25 rules to build up a reliable model for the case of our study. The output of the model is calculated using fuzzy logic toolbox of MATLAB software. And tabled in Table 4.

In order to verify the proposed FIS besed model, the obtained results are compared with Fuzzy-TOPSIS method developed by Chen et al. [2]. In this approach, customer preference are expressed with fuzzy numbers. The final weights of criteria weights for our sample are set to be 0.394, 0.277, 0.257, and 0.106 for unit price, delivery time, quality and past performance respectively (The rules of FIS are also constructed considering these preferences). Using the normalized values of Table 2 and multiplying them by their corresponding weights are results in weighted performance matrix. Then, applying the TOPSIS procedure step by step companies are ranked based on their closeness to the ideal solution as shown in Table 5. The closest candidate to the positive ideal solution has higher partnership chance.

Regarding Table 6 the ranking lists proposed by these two methods are same and both choose Company E as winner. However, their overall chances are not equal. There are two main reasons which explains these differences. First, for constructing FIS based model different types of functions; constant, linear and nonlinear are used. This feature makes the model to be more sensitive to changes in specific domains. Moreover, for criteria such as delivery time and quality which have the predefined "acceptable domain" all of the values within this domain are identically same and does not affect the overall score of candidate. These are the facts neglected in Fuzzy-TOPSIS approach.

Table 3. Set of fuzzy rules.

IF							THEN
Price		Delivery time		Quality		Past perfor-mance	Partnership chance
-		late	OR	less than required		-	low
inexpensive	AND	too early	And	satisfactory	AND	poor	medium
inexpensive	AND	too early	And	satisfactory	AND	good	medium
inexpensive	AND	too early	And	more than required	AND	poor	medium
inexpensive	AND	too early	And	more than required	AND	good	medium
inexpensive	AND	favorable	And	satisfactory	AND	poor	high
inexpensive	AND	favorable	And	satisfactory	AND	good	high
inexpensive	AND	favorable	And	more than required	AND	poor	high
inexpensive	AND	favorable	And	more than required	AND	good	high
average	AND	too early	And	satisfactory	AND	poor	low
average	AND	too early	And	satisfactory	AND	good	low
average	AND	too early	And	more than required	AND	poor	low
average	AND	too early	And	more than required	AND	good	medium
average	AND	favorable	And	satisfactory	AND	poor	medium
average	AND	favorable	And	satisfactory	AND	good	medium
average	AND	favorable	And	more than required	AND	poor	high
average	AND	favorable	And	more than required	AND	good	high
expensive	AND	too early	And	satisfactory	AND	poor	low
expensive	AND	too early	And	satisfactory	AND	good	medium
expensive	AND	too early	And	more than required	AND	poor	low
expensive	AND	too early	And	more than required	AND	good	medium
expensive	AND	favorable	And	satisfactory	AND	poor	low
expensive	AND	favorable	And	satisfactory	AND	good	medium
expensive	AND	favorable	And	more than required	AND	poor	low
expensive	AND	favorable	And	more than required	AND	good	medium

Table 4. Companies partnership chance based on FIS.

Company name	Partnership chance (%)
Company A	60.69
Company B	55.42
Company C	47.73
Company D	58.85
Company E	60.88

Table 5. Companies partnership chance based on fuzzy TOPSIS.

Enterprise name	Distance from positive ideal solution	Distance from negative ideal solution	Closeness (%)
Company A	0.043	0.049	53.3
Company B	0.054	0.041	43.0
Company C	0.066	0.021	24.0
Company D	0.031	0.031	50.0
Company E	0.012	0.072	85.8

Table 6. Companies ranking based on FIS and fuzzy-TOPSIS methods.

Rank	FIS Ranking list	Fuzzy TOPSIS Ranking list
1st	Company E	Company E
2nd	Company A	Company A
3rd	Company D	Company D
4th	Company B	Company B
5th	Company C	Company C

5   Conclusions

In this study, a FIS based model is proposed for formation of virtual enterprise. The developed approach provides an effective tool for ranking the enterprises with respect to both quantitative and qualitative criteria and selecting the best partner to participate in virtual enterprise.

Unless many other techniques in literature, specific membership functions are defined for each criterion regarding their characteristics in order to acquire more reliable outcomes. Besides, the other strength of this method is; it provides a flexible model to change the policies in a way decision maker prefers. Not only fuzzy rules can be edited considering customer preferences, but also membership functions of four inputs and output can be modified corresponding to bidding properties. In contrast, models reliability is highly dependent on establishing reasonable fuzzy rules. Thus, the way to get more accurate results and enhanced models trustworthy is to define precise fuzzy rules by consulting experienced industrial experts.

This study is a preliminary validation of the model for further implementations in industry. The verified model will be implemented in partner selection process of forthcoming VE platform which is going to be established in OSTIM organized industrial park in Ankara.

Acknowledgments. This study is being funded by SAN-TEZ project No. 00979.stz. 2011-12 of Turkish Ministry of Science, Technology and Industry. Authors are sincerely thankful for continuous support of OSTIM Industrial Park management.

References

1. Bevilacqua, M., Petroni, A.: From traditional purchasing to supplier management; a fuzzy logic based approach to supplier selection. Int. J. Logist., Res. Appl. **5**, 235–255 (2010)
2. Chen, T.-Y., Tsao, C.-Y.: The interval-valued fuzzy TOPSIS method and experimental analysis. Fuzzy Sets Syst. **159**(11), 1410–1428 (2008)
3. Fuqing, Z., Yi, H., Dongmei, Y.: A multi-objective optimization model of the partner selection problem in a virtual enterprise and its solution with genetic algorithms. Int. J. Adv. Manuf. Technol. **28**(11–12), 1246–1253 (2005)
4. Huang, X., Wong, Y., Wang, J.: A two-stage manufacturing partner selection framework for virtual enterprises. Int. J. Comput. Integr. Manuf. **17**(4), 294–304 (2004)
5. Mamdani, E., Assilian, S.: An experiment in linguistic synthesis with fuzzy-logic-controller. Int. J. Man-Mach. Stud. **7**, 1–13 (1975)
6. Mikhailov, L.: Fuzzy analytical approach to partnership selection in formation of virtual enterprises. Int. J. Manag. Sci. **30**, 393–401 (2002)
7. Sari, B., Sen, T., Kilic, S.E.: AHP model for the selection of partner companies in virtual enterprises. Int. J. Adv. Manuf. Technol. **38**(3–4), 367–376 (2007)
8. Shing, J., Jang, R.: Adaptive-network-based fuzzy inference system. Trans. Syst., Man Cybern. **23**, 665–685 (1993)
9. Ye, F.: An extended TOPSIS method with interval-valued intuitionistic fuzzy numbers for virtual enterprise partner selection. Expert Syst. Appl. **37**(10), 7050–7055 (2010)
10. Camarinha-Matos, L., Afsarmanesh, H.: The virtual enterprise concept. In: Infrastructure for Virtual Enterprises Networking Industrial Enterprises, pp. 3–4. Kluwer Academic Publishers, London (1999)
11. Zadeh, L.A.: Fuzzy sets. In: Information and Control, pp. 338–353 (1965)
12. Nikghadam, S., Kharrati Shishvan, H., Khanmohammadi, S.: Minimizing earliness and tardiness costs in job-shop scheduling problems considering job due dates. In: Proceedings of AIPE, pp. 181–184. IEEE Press, Kuala lumpur (2011)

Applications

Gasoline Shipper's Problem in Rail Transportation Contract in Thailand

Kannapha Amaruchkul[✉]

Graduate Program in Logistics Management, School of Applied Statistics,
National Institute of Development Administration (NIDA), Bangkok, Thailand
kamaruchkul@gmail.com

Abstract. A long-term contract between a Thai railroad company and
a shipper who wants to transport gasoline daily is studied. The contract
specifies an upfront payment for reserving bogies and a per-container
freight rate. The shipper has to determine the number of bogies to be
reserved upfront at the beginning of the contract, before the random
daily demand is materialized. The shipper can also use a trucking company to transport the excess demand, but its freight rate is higher. We
show that the expected total cost is unimodal under a relatively non-
restrictive condition, and we derive an optimal solution. We also provide
a sensitivity analysis with respect to the change in the contract parameter. Our numerical example shows that the additional cost, if demand
uncertainty is ignored, is large.

Keywords: Stochastic model applications · Rail freight · Logistics management

1 Introduction

Global demand for gasoline and diesel continue to grow over the next few decades
[9]. Population growth is one of the key drivers in growing demand for transportation and motor gasoline consumption [4]. The supply chain for fuel starts
from crude oil sources and ends at gas stations, passing through refinery factories
and storages (e.g., tank farms, depots, terminals, and so on). From the distribution centers to gas stations, gasoline and diesel can be transported through
road, rail, or pipeline if exists.

Pipeline is environment-friendly and always available all year round 24-7-365
with high service level at a relatively low cost. Products are delivered on time
via pipelines, because the flows can be continuously monitored and controlled by
a computer, and they are not affected by weather or climate conditions. Spills
or product losses are minimal, compared to trucks or trains. Accidents involving
trucks or trains with petroleum products are fairly common. Nevertheless, in
some developing countries, e.g., Thailand, pipelines for refined products are not
extensive and cover only a limited geographic area. Rail is better for the environment, produces less greenhouse emission gas, and the transportation cost

© Springer International Publishing Switzerland 2015
D. de Werra et al. (Eds.): ICORES 2015, CCIS 577, pp. 119–135, 2015.
DOI: 10.1007/978-3-319-27680-9_8

is usually lower. On the other hand, road usually allows faster speed, better delivery consistency, and less product losses.

In this article, pipeline is not considered. Gasoline and diesel are transported among regional depots by road or rail. A rail company serves both passengers and cargo shippers/freighters. Bogies (under-carriage assemblies commonly referred to as "trucks" in US) are needed for both cargo tanks and passenger rail cars. When there are not enough bogies for both, the rail company often allocates bogies first to the passenger cars and then box cars for general commodities or tanks for liquid and gas.

To alleviate the problem of bogie availability, some petroleum companies enter a long-term contract with the railroad company. The long-term contract we study is found in practice between the State Railway of Thailand and one of the biggest petroleum companies in Thailand. The contract specifies an upfront payment for reserving bogies and a per-liter freight rate. For each day during the contract period, the number of bogies provided by the railroad company is at most that reserved at the beginning of the contract period. If daily demand is greater than total capacity of reserved bogies, then the excess demand is handled by a trucking company. A per-liter charge by the trucking company is typically higher. The shipper needs to decide how many bogies to reserve before knowing actual daily demand. If the shipper reserves too many bogies in advance, then it would pay a large upfront payment and might end up with low utilization of those reserved bogies, especially when daily demands are not very large. If the shipper reserves not enough bogies upfront, then it might have to pay high transportation cost to the trucking company when daily demands are larger than the total reserved capacity. There is a clear trade-off between too little or too much reserving upfront.

The shipper's problem is similar to the newsvendor (single-period inventory) problem, in which an order quantity must be determined prior to the start of the selling season. In both ours and the newsvendor problem, a fixed quantity is committed before random demand is materialized. Reviews of the newsvendor model can be found in e.g., [5, 11]. Standard textbooks in operations research/management science also discuss a newsvendor problem; e.g., Chap. 10 in [12], Chap. 11 in [3] and Chap. 5 in [8]. In the standard newsvendor model, the total expected cost is shown to be convex, whereas ours is not convex, because the daily transportation rail cost is based on the number of containers actually used, not the demand actually served, as in the standard newsvendor model.

A newsvendor model, which considers standard-sized containers is studied in e.g., [10, 13]. Quantity discount pricing is offered to the shipper. In these papers, if a larger container is used, the per-unit rate is cheaper. Nevertheless, their transportation costs are based on the volume actually shipped, not the number of containers as in ours. Although we do not have quantity discount as in theirs, we have a fixed upfront payment and variable transportation costs. They do not have an upfront payment or a secondary transportation option.

Our contract scheme is related to returns policies or buyback contracts in the newsvendor setting. In our model, the total payment from the shipper to the rail company consists of two parts, namely the upfront payment proportional to the number of reserved bogies, and the variable payment proportional to the actual number of bogies actually used. In the newsvendor model in which a supplier and a buyer enter into a buyback contract, the buyer who places an order quantity of x pays wx to the supplier, where w is the per-unit wholesale price. After demand D materializes, the supplier buys back all unsold units from the buyer at a per-unit buyback price b. The net payment from the buyer to the supplier is

$$wx - b(x - D)^+ = (w - b)x + b\min(x, D)$$

where $(t)^+ = \max(t, 0)$ denotes the positive part of a real number t. The payment under the buyback contract can be viewed as two parts, namely the "upfront" payment, $(w-b)x$, proportional to the committed order quantity and the variable payment, $b\min(x, D)$, proportional to the actual sales. The order quantity in the newsvendor model is analogous to the number of reserved bogies in ours, and the actual sales to the actual number of bogies actually used. Literature on buyback contracts in the newsvendor setting is extensive; see reviews in [2,7]. Ours differs from the buyback contract, because our variable payment is not linear on the actual volume shipped via rail but on the actual number of bogies used. For each realization of demand $D = d$, the variable payment $b\min(x, d)$ in the buyback contract is continuous piecewise linear function in d, whereas our payment is not linear, not continuous in d and has some jumps.

Demands in the newsvendor model and ours are perishable. In our model, demand to transport gasoline must be met on daily basis. The rail company can segment customers, e.g., by freight types. Different freight types can be changed at different prices. The rail company is interested in maximizing revenue because variable costs are small, compared to the fixed sunk cost of acquiring rail cars. Rail freight is a prime candidate for perishable-asset revenue management (RM) techniques. However, papers on railway RM are quite limited, compared to those in "traditional" RM industries, e.g., airline, hotel and car rental. Railway RM papers include, e.g., [1,6].

The rest of the paper is organized as follows: Sects. 1 and 2 give an introduction and a formulation of the problem. We provide an analysis and a numerical example in Sects. 3 and 4 respectively. Section 5 contains a short summary and a few future research directions.

2 Formulation

Throughout this article, let $\mathbb{Z}_+$ denote the set of nonnegative integers, and $\mathbb{R}_+$ the set of nonnegative real numbers.

Consider a shipper who needs to transport fuel (e.g., diesel and gasoline) daily using either road or rail. Let D_i be a random demand (volume in liters) for transport on day i for each $i = 1, 2, \ldots, n$, where n is the length of the planning horizon. Prior to the start of the planning horizon, the shipper and the rail company establish a long-term contract: The rail company guarantees to provide up to $y \in \mathbb{Z}_+$ bogies on each day throughout the planning horizon, and the shipper pays an upfront of $\tilde{f}y$ where $\tilde{f}$ is the per-bogie upfront fee. The upfront payment is collected at the beginning of the planning horizon.

Throughout the planning horizon, the shipper also pays an additional transportation cost, which is linearly proportional to the number of tanks actually used on that day. Let κ be the capacity of the tank (in liters). For day $i = 1, 2, \ldots, n$, the number of tanks actually used by the shipper is:

$$Z_i = \min(y, \lceil D_i/\kappa \rceil) \tag{1}$$

$$= \begin{cases} y & \text{if } y \leq \lceil D_i/\kappa \rceil \\ \lceil D_i/\kappa \rceil & \text{if } y > \lceil D_i/\kappa \rceil \end{cases} \tag{2}$$

$$= \begin{cases} y & \text{if } D_i \geq \kappa y \\ \lceil D_i/\kappa \rceil & \text{if } D_i < \kappa y. \end{cases} \tag{3}$$

To ship total volume of D_i liters, we need $\lceil D_i/\kappa \rceil$ tanks. (We divide the total demand by the tank capacity and round up to its ceiling, because the number of tanks has to be a nonnegative integer.) However, the rail company provides up to the number of reserved bogies y. One tank needs one bogie. Hence, the number of tanks actually used is the minimum of these two quantities. Detailed explanation is as follows: Suppose the demand is larger than the total capacity of reserved bogies ($D_i > \kappa y$). The number of reserved bogie is less than or equal to the number of tanks actually required to accommodate all demand ($y \leq \lceil D_i/\kappa \rceil$). All reserved bogies are used; this corresponds to the upper case in (2) and (3). On the other hand, suppose that the demand is less than the capacity of reserved bogies ($D_i < \kappa y$). We do not need to use all bogies ($y > \lceil D_i/\kappa \rceil$), and only $\lceil D_i/\kappa \rceil$ is actually needed to carry all demand D_i. This corresponds to the lower case in (2) and (3).

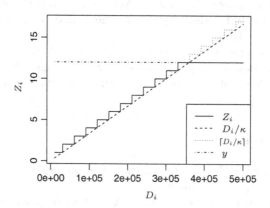

Fig. 1. Number of bogies actually used $Z_i = \min(y, \lceil D_i/\kappa \rceil)$ versus demand, D_i.

Figure 1 shows the number of bogies actually used, when demand ranges from 0 up to 500000 given that capacity is $\kappa = 30000$ and the number of reserved bogies is $y = 12$ (shown in the horizontal dashed-dotted line). The number of bogies needed for all demand is $\lceil D_i/\kappa \rceil$, the stepwise nondecreasing function shown in the dotted line. (The quantity inside the ceiling D_i/κ is a linear increasing function shown in the dashed line.) The solid line in Fig. 1 is the number of bogies actually used, $Z_i = \min(y, \lceil D_i/\kappa \rceil)$ as in (1). (In fact, two bogies are often fitted to each carriage train at the two ends. By "one bogie," we actually mean one pair of bogies. Furthermore, we do not consider a double-stack car/well wagon.)

The additional payment of gZ_i is transferred to the rail company, where the per-liter rail contract rate is r, and the per-tank transportation fee is $g = r\kappa$. If demand exceeds the total volume of tanks initially reserved upfront, κy, then the excess $(D_i - \kappa y)^+$ is transported by the trucking company at a per-liter rate t, where $0 < r < t$. The expected total transportation cost incurred by the shipper is

$$\psi(y) = \tilde{f}y + E\left\{ \sum_{i=1}^{n} \left[g\min(y, \lceil D_i/\kappa \rceil) + t(D_i - \kappa y)^+ \right] \right\}$$

$$= E\left\{ \sum_{i=1}^{n} \left[fy + g\min(y, \lceil D_i/\kappa \rceil) + t(D_i - \kappa y)^+ \right] \right\} \qquad (4)$$

where $f = \tilde{f}/n$ the upfront payment fee per day. Inside the square brackets in (4), the first term is the fixed cost of reserving y bogies, whereas the second and third terms are the random transportation costs by a train and a truck, respectively. Note that (4) is also valid at $y = 0$: If we do not use rail and use only truck, then the expected cost is $\psi(0) = t\sum_{i=1}^{n} E[D_i]$, the per-unit truck cost times the expected total demand.

The shipper's problem is to determine $y \in \mathbb{Z}_+$ the number of bogies to reserve upfront at the beginning of the planning horizon. The trade-offs are obvious: The shipper must commit y before daily demand is materialized. If y is larger than the number of bogies actually needed on that day Z_i, no refund in provided for unused bogies, and the shipper pays too much upfront. On the other hand, if $y < Z_i$, the excess volume needs to be transported by truck, whose per-liter freight rate is more expensive. A firm commitment is hedged against demand uncertainty; this is similar to the newsvendor problem.

Despite some similarity, our payment does differ from that in the newsvendor model. In the standard newsvendor problem, the objective function is based on the demand actually served, $\min(x, D)$, where D is a random demand and x is the order quantity. If the daily payment to the rail company were linearly proportional to daily volume actually shipped, then the standard newsvendor problem would be applied directly. Nevertheless, the daily payment to the rail company is linearly proportion to the total volume of tanks actually used, $\kappa\min(y, \lceil D_i/\kappa \rceil)$, which is greater than or equal to the total volume actually shipped, $\min(\kappa y, D_i)$. Then,

$$g\min(y, \lceil D_i/\kappa \rceil) = r\{\kappa\min(y, \lceil D_i/\kappa \rceil)\}$$
$$\geq r\kappa\min(y, D_i/\kappa) = r[\min(\kappa y, D_i)]. \qquad (5)$$

The left-hand-side (LHS) can be viewed as the freight rate times the total volume of tanks, whereas the right-hand-side (RHS) is the freight rate times the total volume actually carried in the tanks. The total volume of tanks (LHS), the term inside the curly brackets, is at least the total volume actually carried (RHS), the term inside the square brackets. Our scheme implies higher payment to the rail company, compared to other schemes based on the volume actually shipped. In the analysis below, we will show that, because of this payment scheme, the expected cost is not convex, whereas that in the newsvendor model is convex.

3 Analysis

Assume that the vector of daily demands $D_1, D_2, \ldots, D_n$ are independent and identically distributed $\mathbb{R}_+$-valued random variables. Let F be the common distribution and D the random variable with such distribution, i.e., the random daily demand. Let $\bar{F}$ be the complementary cumulative distribution function; i.e., $\bar{F}(t) = P(D > t) = 1 - P(D \leq t) = 1 - F(t)$. Let F^{-1} denote the quantile function. Define the first-order loss function of D as $L(x) = E[(D - x)^+] = \int_x^\infty \bar{F}(u)du$. Also define the limited expected value as $\text{LEV}(x) = E[\min(x, D)] = \int_0^x \bar{F}(u)du$. For well-known continuous distributions (e.g., lognormal, gamma, Weibull, beta), the formulas for the limited expected value and the loss function are readily available. Furthermore, if an expression for $L(x) = E[(D - x)^+]$ is available, then we can easily obtain a formula for $\text{LEV}(x) = E[\min(x, D)]$ by using $E[\min(x, D)] = E[D] - E[(D - x)^+]$.

Define the expected daily cost, if the number of reserved bogies is y, as follows:

$$\pi(y) = fy + gE[\min(y, \lceil D/\kappa \rceil)] + tE[(D - \kappa y)^+] \tag{6}$$

$$= fy + g \sum_{i=0}^{y-1} \bar{F}(i\kappa) + t \int_{\kappa y}^\infty \bar{F}(u)du. \tag{7}$$

Then, the total expected cost over n day is

$$\psi(y) = n\pi(y) \tag{8}$$

since the expectation of the sum of random variables is the sum of the expected values. [Derivation of (7) is provided in Appendix.] Henceforth, we focus on the expected daily cost. Note that expression (7) provides an easy way to evaluate $\pi(y)$: The last term $E[(D-x)^+] = \int_x^\infty \bar{F}(u)du$ where $x = \kappa y$ is the first-order loss function of D.

Proposition 1 characterizes an optimal solution $y^* = \text{argmin}\, \pi(y)$ when demand is deterministic; specifically, $P(D = d) = 1$. Proofs of all propositions are given in Appendix.

Proposition 1. *Suppose $f \leq (t - r)\kappa$.*

1. If $d < (f+g)/t$, $y^* = 0$, and only truck is used.
2. If $(f+g)/t \leq d \leq \kappa$, $y^* = 1$, one bogie is used.
3. If $d > \kappa$, both train and truck are used, and

$$y^* = \mathrm{argmin}\{\pi(\lfloor d/\kappa \rfloor), \pi(\lceil d/\kappa \rceil)\}.$$

Suppose $f > (t-r)\kappa$. Then, $y^* = 0$ and only truck is used.

The results in Proposition 1 make sense economically. If the fixed upfront is very large [i.e., $f > (t-r)\kappa$], then we use only truck. On the other hand, suppose that the fixed upfront is not that large. Then we have to take into account the size of demand. If demand is smaller than the cutoff $(f+g)/t$, we use only truck to avoid paying upfront. If demand is larger than the cutoff and less than the bogie capacity, then it is optimal to reserve only one bogie, and no truck is needed, because all demands can be put into one rail tank. If demand is larger than the bogie capacity, then we use both truck and train. The cost function is minimized at either of the two kinks, $\lfloor d/\kappa \rfloor$ or $\lceil d/\kappa \rceil$. The optimal number of bogies is obtained by dividing demand by capacity and rounding down if $\pi(\lfloor d/\kappa \rfloor) < \pi(\lceil d/\kappa \rceil)$ or rounding up otherwise.

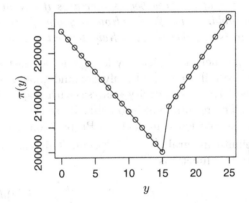

Fig. 2. Expected daily cost $\pi(y)$ is not convex but unimodal on $\mathbb{Z}_+$.

Figure 2 shows $\pi(y)$ with constant demand $P(D = d) = 1$ where $d = 458000$, $\kappa = 30000$, $r = 0.3691$, $t = 0.49$, $f = 2000$. There are two kinks at $\lfloor d/\kappa \rfloor = 15$ and $\lceil d/\kappa \rceil = 16$. Here, the optimal number of bogies is obtained by dividing demand by capacity and rounding down, since $\pi(\lfloor d/\kappa \rfloor) < \pi(\lceil d/\kappa \rceil)$. Note that the cost is not convex. Nevertheless, it is unimodal; thus, we can derive an optimality condition by

$$\mathrm{argmin}\, \pi(y) = \mathrm{argmin}_{y \in \mathbb{Z}_+}\{\Delta(y) \geq 0\} \qquad (9)$$

where $\Delta(y) = \pi(y+1) - \pi(y)$ denote the forward difference.

Henceforth, we consider the problem with random demand. In Proposition 2 we identify a condition under which the expected cost function is unimodal so the optimality condition (9) holds, and we provide a closed-form optimal expression for the number of reserved bogies, which minimizes the expected daily cost. Let ζ denote a density function of F.

Proposition 2. *Assume that*

$$\frac{\kappa\zeta(\kappa y)}{F(\kappa(y+1)) - F(\kappa y)} \leq \frac{t}{r} \tag{10}$$

for all $y \in \mathbb{Z}_+$.

1. *The expected cost $\pi(y)$ is unimodal and attains its minimum at*

$$y^* = \operatorname{argmin}_{y \in \mathbb{Z}_+}\{\Delta(y) \geq 0\}$$

   where

$$\Delta(y) = f + r\kappa\bar{F}(\kappa y) - t\int_{\kappa y}^{\kappa(y+1)} \bar{F}(u)du. \tag{11}$$

   Furthermore, if $f \geq \kappa(t-r)$, then $\pi(y)$ attains its minimum at $y^* = 0$.
2. *The optimal number of reserved bogies increases if one or more of the following conditions hold: (i) the fixed upfront payment f decreases; (ii) the rail freight rate r decreases; (iii) the truck freight rate t increases.*

If the fixed upfront payment is sufficiently large, e.g., at least $\kappa(t-r)$, then the shipper should not use rail service; all daily demand should be accommodated by truck. Part 2 also gives a sensitivity analysis with respect to a change in a contract parameter. The results make economically sense.

To find an optimal solution using (11) in Proposition 2, we need to be able to evaluate the definite integral $\int_{\kappa y}^{\kappa(y+1)} \bar{F}(u)du$. This can be written as the difference of the two loss functions

$$\int_{\kappa y}^{\kappa(y+1)} \bar{F}(u)du = \int_{\kappa y}^{\infty} \bar{F}(u)du - \int_{\kappa(y+1)}^{\infty} \bar{F}(u)du$$

$$= L(\kappa y) - L(\kappa(y+1)).$$

Alternatively, it can be computed as the difference of the two limited expected values

$$\int_{\kappa y}^{\kappa(y+1)} \bar{F}(u)du = \int_{0}^{\kappa(y+1)} \bar{F}(u)du - \int_{0}^{\kappa y} \bar{F}(u)du$$

$$= \operatorname{LEV}(\kappa(y+1)) - \operatorname{LEV}(\kappa y).$$

As previously mentioned, a closed-form expression for either loss function or limited expected value can be found in standard probability and statistics textbooks. A numerical example, which describes how to find an optimal solution, is given in the next section.

Although the expected cost $\pi(y)$ is not convex, it is unimodal, and an optimal solution can be found analytically, when condition (10) holds. We argue that this condition is not very restrictive. Recall that the cumulative distribution function is the area under the density curve: $F(t) = \int_0^t \zeta(u)du$. Then,

$$F(\kappa(y+1)) - F(\kappa y) \approx \kappa\zeta(\kappa y). \tag{12}$$

In (12), the LHS is the area under the density curve from $t_1 = \kappa y$ to $t_2 = \kappa(y+1)$, whereas the RHS is the area of the rectangle with height equal to density at t_1, $\zeta(\kappa y)$, and width equal to the length of interval $[t_1, t_2]$, κ. With this approximation, condition (10) becomes

$$1 \approx \frac{\kappa\zeta(\kappa y)}{F(\kappa(y+1)) - F(\kappa y)} \leq \frac{t}{r}.$$

In our model, the truck rate is strictly greater than the rail rate, $t > r$; so condition (10) usually holds.

We now turn our attention to a heuristic solution. We approximate the expected daily cost by removing the ceiling function in (6):

$$\nu(y) = fy + gE[\min(y, D/\kappa)] + tE[(D - \kappa y)^+]$$
$$= fy + rE[\min(\kappa y, D)] + tE[(D - \kappa y)^+].$$

The approximated cost $\nu(y)$ differs from the exact cost $\pi(y)$ only on the second term: In $\nu(y)$, the cost charged by the rail company is based on the total volume actually shipped, whereas in $\pi(y)$ it is based on the total volume of tanks actually used; see (5). Clearly, the approximated cost is a lower bound on the exact cost, by the definition of the ceiling function. The approximated cost $\nu(y)$ is continuously differentiable, and a one-line formula for a minimizer is derived in Proposition 3.

Proposition 3. *If $f < \kappa(t-r)$, then the approximated cost $\nu(y)$ is convex and attains its minimum at*

$$y_a = \frac{1}{\kappa}F^{-1}\left(1 - \frac{f/\kappa}{t-r}\right). \tag{13}$$

We refer to $1 - f/[\kappa(t-r)]$ as the *critical ratio*. Recall that in the newsvendor model with overage c_o and underage c_u, the optimal order quantity is $F^{-1}(c_u/(c_u + c_o))$ where F is the demand distribution, and $c_u/(c_u + c_o)$ is the so-called critical ratio. In analogous, the overage is f/κ, the per-liter cost of reserving too much and some bogies are not used. The underage occurs if we do not reserve enough bogies and need to use both road and rail; so the per-liter underage is $t - r - f/\kappa$. The underage can be viewed as the incremental per-liter cost if truck is used. Specifically, it is $t - (r + f/\kappa)$, the per-liter cost by truck t minus the per-liter cost by rail, which is the sum of the per-liter freight rate and the per-liter upfront $(r + f/\kappa)$.

Our heuristic solution is to reserve

$$y_a^* = \operatorname{argmin}\{\pi(\lceil y_a \rceil), \pi(\lfloor y_a \rfloor)\}.$$

Our heuristic solution requires evaluation of the expected daily cost function $\pi(y)$ and compares the expected values at the floor and ceiling of y_a given in Proposition 3.

In practice, the per-liter rates by train and truck, r and t, may depend on daily crude oil or gas price; they may be random when the contract is signed, and their values are realized daily. Then, the expected total cost (4) becomes

$$E\Big\{ \sum_{i=1}^{n} \big[fy + \kappa R_i \min(y, \lceil D_i/\kappa \rceil) + T_i(D_i - \kappa y)^+ \big] \Big\}$$

where R_i and T_i are per-liter rates by train and truck on day i, respectively. We further assume that the per-liter rates are independent and identically distributed and that they are independent of demands. Our formulation and analysis remain valid, if we now interpret $r = E[R_1]$ and $t = E[T_1]$ as the expected per-liter rates by train and truck, respectively.

4   Numerical Example

Consider one of the biggest petroleum companies in Thailand, who wants to transport hi-speed diesel from Saraburee province in the central region to Phare province in the northern region. Two modes of transportation are available: (1) Road, service offered by a 3PL trucking company (2) Rail from Baan Pok Pak station in Saraburee to Denchai station in Phare, service offered by the State Railway of Thailand. The distance is about 500 km. Either truck or train can make a round trip within a single day.

The capacity of a rail tank is $\kappa = 33000$ liters. The rail per-liter rate is $r = 0.3169$ Thai Baht (THB), the per-tank cost is $g = r\kappa = 10457.7$ THB, and the truck per-liter rate is $t = 0.49$ THB, which is 0.1731 THB/liter higher than the rail rate. The daily upfront payment is $f = 92$ THB; i.e., the annual per-bogie upfront fee is $\tilde{f} = 28520$ THB, where we assume that there are $n = 310$ operating days in a year.

The 310-day demands in year 2013 are collected. We divide the data into two sets as follows: The first set consists of daily demands in quarters 1 through 3, and the second set consists of daily demands in quarter 4. We use the first set to build the stochastic model and obtain the corresponding decision on the number of reserved bogies. We use the second set as a *holdout sample* to evaluate performance of the proposed solution.

Proposed Solutions

To obtain the optimal number of reserved bogies, we use Proposition 2. First, we need to estimate the demand distribution from the historical data. We provide some classical descriptive statistics as follows: Minimum is 44909; maximum

1123415; median 363139; mean 373772.7; estimated standard deviation 165346.5; estimated skewness 0.6292; estimated kurtosis 4.1345. The coefficient of variation is 0.44; in words, the standard deviation is a bit less than half of the mean. The mean is greater than the median, and skewness is positive. We consider common right-skewed distributions, namely Weibull, lognormal and gamma. We use maximum likelihood estimation (MLE) for the parameters of Weibull and lognormal. For the gamma distribution, MLE fails to converge, and the method of moments is used instead. Table 1 shows the parameter estimates of the three distributions and the p-values of the Kolmogorov-Smirnov (K-S) goodness of fit test. Using the 10 % significance level, we choose to build models based on the Weibull and gamma distributions. There is sufficient evidence to reject the null hypothesis that data comes from the lognormal distribution with parameters $(12.7188, 0.5082)$ since the p-value $0.0535 \leq 0.10$; we do not use the lognormal. The histogram is shown in Fig. 3, and the corresponding densities are superimposed over it.

Table 1. Result from fitting distribution.

Distribution	Parameter 1	Parameter 2	The p-value
Weibull	2.402162	421709.4	0.9217
Gamma	5.132078	0.0000137	0.4803
Lognormal	12.718847	0.5081968	0.0535

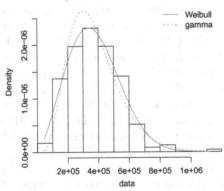

Fig. 3. Histogram of daily demand.

We present a detailed solution for the Weibull distribution. The solution for the gamma distribution can be obtained in a similar manner. Proposition 2 assumes that condition (10) holds for all $y \in \mathbb{Z}_+$. For the given daily demand distribution, we choose the maximum number of reserved bogies to be $y^m = 39$; this y^m is sufficient to accommodate daily demand roughly 99.99 percent of the time. We can verify that condition (10) holds for each $y = 0, 1, 2, \ldots, y^m$. [Specifically, the maximum of the LHS in (10) for each $y = 0, 1, 2, \ldots, y^m$ is 1.499116, whereas the value of the RHS is $t/r = 1.5462$.] Then, we use (11) to find an optimal solution. The difference $\Delta(y)$ is evaluated for each $y = 1, 2, \ldots, y^m$. The optimal solution given in Proposition 2 requires the shipper to reserve $y^* = \mathrm{argmin}\{\Delta(y) \geq 0\} = 22$ bogies, denoted this solution approach as OPT. To find a heuristic solution, we compute the critical ratio given in (13) $1 - f/[\kappa(t-r)] = 0.9839$; the corresponding quantile divided by the capacity is $y_a = 23.1$. Since $\pi(23) = 125920 < 125968.4 = \pi(24)$, from our heuristic the number of reserved bogies $y_a^* = 23$, denoted as H3. At this particular problem instance, the heuristic and optimal solutions are differed by just one bogie. Table 2 presents the solutions from the two distributions.

Table 2. Solutions from the two distributions.

Distribution	y^m	maximum of LHS in (10)	OPT	H3
Weibull	39	1.499116	22	23
Gamma	53	1.199536	23	24

Suppose that one completely ignores demand uncertainty; i.e., demand is assumed to be deterministic. Then two simple heuristics can be developed. We compute the average demand $\bar{d} = 373772.7$ and assume that demand is constant and equal to this value. Proposition 1 gives the "optimal" number of reserved bogies when demand is assumed to be constant. Here, the upfront is not too large $[f \leq (t - r)\kappa = 5712.3]$, and demand is larger than capacity ($\bar{d} > \kappa$): It follows from part 3 in Proposition 1 that the number of reserved bogies is $y_1 = 11$, denoted as H1. Suppose we further ignore all cost parameters. Then, one could estimate the number of reserved bogies by dividing the daily mean demand by the tank capacity and rounding up, $y_2 = \lceil \bar{d}/\kappa \rceil = \lceil 11.3 \rceil = 12$, denoted as H2. Table 3 shows all the solutions we obtain. Note that the numbers of reserved bogies when demand uncertainty is ignored in H1 and H2 are much less than those when randomness is explicitly incorporated in H3 and OPT.

Throughout our numerical example, the LEV function is used extensively to evaluate $\Delta(y)$ and $\pi(y)$. All calculations and statistical analyses in our numerical example are done in R. For instance, descdist in package fitdistrplus gives descriptive statistics; fitdist fits distribution; ks.test performs the K-S test; pweibull returns the cumulative distribution function of Weibull distribution; Gamma calculates the complete gamma function.

Evaluation of Proposed Solutions

We evaluate the performance of the solution using the holdout sample. Our holdout sample consists of $m = 77$ days. Given that the number of reserved bogies is y, the cost on day i is

$$\psi_i(y) = fy + g \min(y, \lceil d_i/\kappa \rceil) + t(d_i - \kappa y)^+ \qquad (14)$$

where d_i is demand on day $i = 1, 2, \ldots, m$. The first term is the fixed cost, the second the (variable) transportation cost via rail, and the third the truck cost. The cost throughout m day is $\sum_{i=1}^{m} \psi_i(y)$. The total yearly cost is estimated as $(n/m) \sum_{i=1}^{m} \psi_i(y)$. Table 3 shows the yearly fixed cost $\tilde{f}y = nfy$, the yearly rail cost (excluding the fixed cost), the yearly truck cost and the total yearly cost for each solution.

Table 3. Yearly costs for each of the four proposed solutions.

No. of bogies	Solution approaches	Yearly Cost (THBs)				
		Rail	Truck	Fixed	Total	Diff
11	H1	32797792	13623410	313720	46734922	3295732
12	H2	34860811	10857447	342240	46060498	2621308
22	OPT-Weibull	42102429	767193	627440	43497062	57872
23	H3-Weibull, OPT-Gamma	42228736	571893	655960	43456589	17399
24	H3-Gamma	42355043	399667	684480	43439190	0

The approaches based on either Weibull or gamma distributions (OPT and H3) perform much better than those when demand uncertainty is ignored (H1 and H2). In other words, the two heuristics based on the assumption of constant demand perform very poorly. The additional cost per year is about three million THBs if demand uncertainty is ignored; see the last column in Table 3. The heuristic solution based on gamma distribution performs best. In each of the two distributions, H3 slightly outperforms OPT. H3 is obtained based on the approximated cost function in Proposition 3. H3 requires much less calculation effort; the one-line formula (13) can be used to find the number of reserved bogies. The numerical example shows that our proposed solutions in Propositions 2–3 perform well on real data.

5   Conclusion

In summary, we formulate a stochastic model in which the shipper determines how many bogies to be reserved before random daily demand is materialized. If the number of reserved bogies is very large, we may end up paying a large upfront fee for unused bogies. On the other hand, if we reserve not enough bogies, we incur a large transportation cost via truck. We want to determine an optimal number of bogies to reserve in order to minimize the expected cost. In the analysis, we show that under certain condition, the expected cost is unimodal and a closed-form optimal solution is derived. An easy-to-implement heuristic solution is also proposed. Some sensitivity analysis is provided: The number of reserved bogies decreases, if the upfront payment increases, or the rail freight rate increases, or the truck freight rate decreases. In the numerical example, we show that the estimated cost using our optimal solution and our heuristic solution are about the same. If we did not take into account of demand uncertainty as in our model, then the number of reserved bogies could differ from the optimal solution by almost 100 % and would result in a significant increase in transportation cost.

Some future research directions are identified below. Recall that, in the analysis, we assume demands are independent and identically distributed. Note that the expected total cost (4) can be written as

$$\psi(y) = E[\sum_{i=1}^{n} \psi_i(y)] = \sum_{i=1}^{n} E[\psi_i(y)] \qquad (15)$$

where $\psi_i(y)$, previously defined in (14), depends on the random demand on day i. Eq. (15) always holds (whether or not demands are independent) since the expectation of the sum of random variables is equal to the sum of the expectations. If demands are not independent but identically distributed, then all of our results remain unchanged, because $\psi(y) = nE[\psi_1(y)]$ as in (8), and we can focus on the expected daily cost, $E[\psi_1(y)] = \pi(y)$ in (6). However, if demands are not identically distributed, then we need to minimize the total expected cost $\sum_{i=1}^{n} E[\psi_i(y)]$: We cannot minimize each term $E[\psi_i(y)]$ separately, because in general argmin $E[\psi_i(y)] \neq$ argmin $E[\psi_j(y)]$ for each $i \neq j$. Cases when demands are nonstationary (i.e., neither independent nor identically distributed) would be an interesting research extension.

Recall that our model considers a contract problem on a single leg in the railroad network. One extension would be to study a contract to transport gasoline over the entire railroad network. Furthermore, for each leg, there may be multiple types of containers, depending on their capacities. The upfront fee depends on the container type, and the per-container charge depends on the container capacity and the distance between an origin and a destination.

Another interesting extension is a railroad's contract design problem. Our model allows the shipper to choose the number of bogies to be reserved. In the contract design problem, we would allow the railroad company to choose the parameters of the contract, e.g., the upfront payment and the per-container freight charge. The contract design problem can be modeled using a game-theoretical framework. We hope to pursue these or related problems in the future.

Acknowledgements. The original problem was materialized after some discussions with one of our part-time master students, Mr. Apichat Gunthathong, who has been working at the petroleum company (in the numerical example) for 12 years. His independent project, a part of requirement for a master degree in logistics management at the school, was related to our model.

Appendix

Derivation of (7). For shorthand, denote $M = \lceil D/\kappa \rceil$. Since D is a nonnegative random variable with distribution F, M is a $\mathbb{Z}_+$-valued random variable. The cumulative distribution function of M is as follows: For any $t \in \mathbb{R}_+$

$$P(M \le t) = P(\lceil D/\kappa \rceil \le \lfloor t \rfloor)$$
$$= P(D/\kappa \le \lfloor t \rfloor)$$
$$= F(\lfloor t \rfloor \kappa).$$

We use the tail-sum formula for expectation:

$$E[\min(t, M)] = \sum_{j=0}^{\infty} P(\min(t, M) > j)$$
$$= \sum_{j=0}^{\lfloor t \rfloor - 1} P(M > j)$$
$$= \sum_{j=0}^{\lfloor t \rfloor - 1} \bar{F}(j\kappa).$$

The finite sum in the second term in (7) is obtained, and the last term is the first-order loss function of D.   □

Proof of Proposition 1

Proof. For shorthand, let

$$\tilde{\pi}(y) = [f - (t - r)\kappa]y + td$$
$$\hat{\pi}(y) = fy + g\lceil d/\kappa \rceil.$$

Given that demand is constant and equal to d, the daily cost (6) becomes

$$\pi(y) = fy + g \min(y, \lceil d/\kappa \rceil) + t(d - \kappa y)^+$$
$$= \begin{cases} fy + gy + t(d - \kappa y) & \text{if } y = 0, 1, 2, \ldots, \lfloor d/\kappa \rfloor \\ fy + g\lceil d/\kappa \rceil & \text{if } y = \lceil d/\kappa \rceil, \ldots \end{cases}$$
$$= \begin{cases} \tilde{\pi}(y) & \text{if } y = 0, 1, 2, \ldots, \lfloor d/\kappa \rfloor \\ \hat{\pi}(y) & \text{if } y = \lceil d/\kappa \rceil, \lceil d/\kappa \rceil + 1, \ldots \end{cases}$$

In particular, if $d < \kappa$,

$$\pi(0) = td$$
$$\pi(y) = fy + g \quad \text{if } y = 1, 2, \ldots$$

Since both $\tilde{\pi}(y)$ and $\hat{\pi}(y)$ are linear, $\pi(y)$ is a piecewise linear function. Note that $\hat{\pi}(y)$ is strictly increasing since $f > 0$. Let $y^* = \text{argmin}\,\pi(y)$.

Suppose $f > (t - r)\kappa$. Both $\tilde{\pi}$ and $\hat{\pi}$ are increasing functions, so $y^* = 0$. In words, if the fixed upfront is sufficiently large, we use only truck and do not reserve any bogies ($y^* = 0$).

Suppose $f \le (t - r)\kappa$. That is, $(f + g)/t \le \kappa$ since $g = r\kappa$.

1. If $d < (f + g)/t$, then in this case $d < \kappa$ and $\pi(0) = td \le (f + g) = \pi(1)$, so $y^* = 0$, and we use only truck.
2. If $(f + g)/t \le d < \kappa$, then $\pi(0) > \pi(1)$, so $y^* = 1$; i.e., we reserve one bogie. Since demand does not exceed capacity, one bogie is enough to accommodate entire demand; we do not need any trucks.
3. If $d \ge \kappa$, then $\pi(y)$ is linearly decreasing up to $\lfloor d/\kappa \rfloor$ and increasing from $\lceil d/\kappa \rceil$. Specifically, the coefficient of y in $\tilde{\pi}(y)$ is negative when $f \le (t-r)\kappa$, so $\tilde{\pi}(y)$ is linearly decreasing and minimized at $\lfloor d/\kappa \rfloor$. Since $\hat{\pi}(y)$ is increasing, it is minimized at $\lceil d/\kappa \rceil$. Thus, the global minimizer is $\text{argmin}\{\hat{\pi}(\lceil d/\kappa \rceil), \tilde{\pi}(\lfloor d/\kappa \rfloor)\}$.

□

Proof of Proposition 2

Proof. From (7) we have that

$$\Delta(y) = f + g\bar{F}(\kappa y) - t \int_{\kappa y}^{\kappa(y+1)} \bar{F}(u)\,du,$$

and

$$\Delta'(y) = -\kappa g\zeta(\kappa y) + \kappa t[F(\kappa(y+1)) - F(\kappa y)].$$

Recall $g = r\kappa$. If (10) holds, then $\Delta'(y) \ge 0$, or equivalently $\Delta(y)$ is increasing; thus, the expected cost function $\pi(y)$ is unimodal. Let $y^* = \text{argmin}_{y \in \mathbb{Z}_+}\{\Delta(y) \ge 0\}$. Then, $\pi(y)$ switches from decreasing to increasing at y^*; hence, $\pi(y)$ attains its minimum at y^*.

Finally, note that

$$\Delta(0) = f + \kappa r - t \int_0^\kappa \bar{F}(u)\,du \ge f - \kappa(t - r)$$

since $0 \le \bar{F}(u) \le 1$. If $f \ge \kappa(t - r)$, $\Delta(0) \ge 0$; $\pi(y)$ is increasing on $\mathbb{Z}_+$ and is minimized at $y^* = 0$. Part 2 follows from (11) and the fact that $\Delta(y)$ is increasing.

□

Proof of Proposition 3

Proof. For convenient let $x = \kappa y$ and define

$$\xi(x) = (f/\kappa)x + rE[\min(x, D)] + tE[(D - x)^+]$$

Then, $\operatorname{argmin} \nu(y) = (\operatorname{argmin} \xi(x))/\kappa$ since $\nu(y) = \xi(x)$. After some simplifications, we get

$$\xi(x) = (f/\kappa)x + rE[D] + (t - r)E[(D - x)^+].$$

The first and second derivatives with respect to x are

$$\xi'(x) = f/\kappa - (t - r)\bar{F}(x)$$
$$\xi''(x) = (t - r)F'(x) = (t - r)\zeta(x) \geq 0$$

respectively. The approximated cost function $\xi(x)$ is convex. If $f < \kappa(t - r)$, then the optimality condition is that $\xi'(x) = 0$; i.e., $x = F^{-1}(1 - f/[\kappa(t - r)])$ and (13) follows immediately.   □

References

1. Armstrong, A., Meissner, J.: Railway revenue management: Overview and models. Working paper, Lancaster University Management School (2010). http://www.meiss.com
2. Cachon, G.: Supply chain coordination with contracts. In: de Kok, A., Graves, S. (eds.) Handbooks in Operations Research and Management Science: Supply Chain Management. Elsevier, Amsterdam (2003)
3. Cachon, G., Terwiesch, C.: Matching Supply with Demand: An Introduction to Operations Management. McGraw-Hill Inc, New York (2009)
4. Energy Information Administration (EIA). Motor gasoline consumption 2008: A historical perspective and short-term projections (2008). www.eia.gov/forecasts/steo/special/pdf/. Accessed 2 August 2014
5. Khouja, M.: The single-period (news-vendor) problem: literature review and suggestions for future research. Omega **27**(5), 537–553 (1999)
6. Kraft, E., Srika, B., Phillips, R.: Revenue management in railroad applications. Transp. Q. **54**(1), 157–176 (2000)
7. Lariviere, M.: Supply chain contracting and coordination with stochastic demand. In: Tayur, S., Ganeshan, R., Magazine, M. (eds.) Quantitative Models for Supply Chain Management. Kluwer Academic Publishers, Boston (1999)
8. Nahmias, S.: Production and Operations Research. McGraw-Hill Inc, New York (2009)
9. Organization of the Petroleum Exporting Countries (OPEC) (2013). World oil outlook 2013. www.opec.org. Accessed 2 August 2014
10. Pantumsinchai, P., Knowles, T.: Standard container size discounts and the single-period inventory problem. Decis. Sci. **22**(3), 612–619 (1991)
11. Qin, Y., Wang, R., Vakharia, A., Chen, Y., Seref, M.: The newsvendor problem: review and directions for future research. Eur. J. Oper. Res. **213**(2), 361–374 (2011)
12. Silver, E., Pyke, D., Peterson, R.: Inventory Management and Production Planning and Scheduling. Wiley, New York (1998)
13. Yin, M., Kim, K.: Quantity discount pricing for container transportation services by shipping lines. Comput. Ind. Eng. **63**(1), 313–322 (2012)

An Optimization Model for Planning Operations in a Meat Packing Plant

Victor M. Albornoz[1], Marcela Gonzalez-Araya[2], Matias C. Gripe[1],
Sara V. Rodriguez[3]([✉]), and E. Juventino Treviño[3]

[1] Departamento de Industrias, UTFSM, Av. Santa Maria, 6400 Santiago, Chile
[2] Departamento de Ingenieria Industrial, Universidad de Talca, Talca, Chile
[3] Facultad de Ingenieria Mecanica y Electrica,
Universidad Autonoma de Nuevo Leon,
San Nicolas de Los Garza, Mexico
sara.rodriguezsn@uanl.edu.mx

Abstract. This paper presents an optimization model for supporting planning operations in a meat packing plant. The production system considers processing raw material (carcasses) by applying cutting patterns in order to meet a given demand of different products. Major decisions in the planning problem include the number of times each cutting pattern is applied on the available carcasses, the total yield per product and its corresponding levels of inventory at each time period; as a result a Mixed Integer Linear Programming Model (MILP) is proposed. The main contribution is to give an optimal decision that maximizes the economic profit and thereby reducing the shortages for products with the highest profits and the inventory for those with low level demand, taking into account the perishability of the products, labour capacities, and the fact that it may exists different types of carcasses with different yields.

Keywords: MILP · Production planning · Meat packing plant · Cutting patterns · Perishability

1 Introduction

Operations Research is one of the most important disciplines that deal with advanced analytical methods for decision making. It is applied to a wide range of problems arising in different areas such as operations management in agriculture and food industry, see [1, 3–5].

In particular, the decision making for planning operations in a meat packing plant has been studied by several authors. To the best of authors knowledge the first contribution was done by [10], who presented a linear programming formulation of a partitioned cutting stock problem applied in the meat industry. The main feature of their formulation is the partitioning of cutting patterns by carcass sections that greatly reduces the number of cutting patterns in the formulation. However, such model was devoted to marketing planning purposes, and consequently lacks of the elements that support the production plan, such as inventories, capacities, demand and time horizon, among others.

© Springer International Publishing Switzerland 2015
D. de Werra et al. (Eds.): ICORES 2015, CCIS 577, pp. 136–146, 2015.
DOI: 10.1007/978-3-319-27680-9_9

Years later [9] presents a contribution for a meat packing plant production plan, using a Mixed Integer Goal Programming formulation to pursue multiple objectives, however such a model does not take into account that in a given batch there may be different types of carcasses, and variations in some parameters. In 2006, [2] presents an integrated system of 45 linear programming models to schedule operations in a real case for Swift & Company, a beef meat packing plant. Several thesis dissertations have also been found in the literature. For instance, [11] develops on-line optimization techniques for determining which cutting pattern to use in each carcass according to both carcass attributes and demand, but in a real time horizon. The paper of [6] presents a linear programming formulation to maximize the value of pork products. In [7] we find a DEA study on some parameters and uses a planning model to determine the levels of pork production by product. Whereas [8] develops a planning production model for a beef meat packing plant but without including multi-periods.

Although the approaches existing in the literature, there exist some gaps that need to be considered, such as the establishment of the production plan considering different types of carcasses. The aim of this paper is to present a mathematical optimization model to support the planning operations in a meat packing plant. In particular, we formulate a linear mixed integer program (MILP) that addresses decisions that indicates how often a cutting pattern is applied on each type of carcasses, in order to keep a certain level of inventory, meeting demand requirements and other storage capacities, and labour time, in each period of time. Section 2 presents the problem statement. Section 3 introduces the mixed integer linear programming formulation. Section 4 summarizes the computational results and finishes with a sensitivity analysis on some stochastic parameters. Section 5 summarizes the main conclusions.

2   Problem Statement

The production planning establishes the resources to be procured (carcasses), the levels of production and the corresponding level of inventory per product. Such decisions depend on how to cut up the carcasses in order to satisfy consumer demand. Several cutting patterns exist in the market; each cutting pattern involves a corresponding set of products, specific rewards and operating costs. It is possible that different cutting patterns share a common product, in other words a specific product can be obtained through different cutting patterns. Nonetheless, the yield per product obtained may be different and related to the cutting pattern. This problem in an isolated way could be tackled through data analysis, but in practical implementation, its interaction with the demand makes it much more complex to solve.

At the operative level the manager must consider not only the yield of products and carcass availability, but also the demand behaviour. Demand behaviour is not constant throughout time, and moreover it is not homogeneous among all pork products; each product has its own level of demand. Such issues are particularly relevant in disassemble problems such as the one involved in the production of pork. A previous data analysis on demand shows there exists a set of pork products (from the entire carcass) with large demand, while other products have small or no market at all. Hence, the major difficulty is to balance the benefits between demand and production, while managing inventories of perishable products. The main concern here is to determine the number of times each cutting pattern is applied to the available carcasses, and the levels of production obtained for the entire list of products. Large variability, uncertainty, perishability, large scale of operations and long lead times are some issues that most pork supply chain managers must face at the operational level.

3   Mathematical Formulation

This section provides a detailed description of the mathematical formulation to support the decision making for planning operations in a meat packing plan. Major decisions include the number of times each cutting pattern must be applied on the available types of carcass, the total yield per product and its corresponding levels of inventory, at each time period. The model assumes that the company is facing shortages, however in order to keep unsatisfied demand as low as possible, a shortage cost is considered in the objective function and whose value is equivalent to the selling price of competitors. In addition to the inventory and shortages decisions, the model also considers overtime as an additional flexibility to face stochastic demand and large variability in carcass traits. Finally, the decision variables include two types of presentations for production and inventory levels: fresh and frozen.

The objective function maximizes the net profit of the producer taking into consideration different constraints to meet demand, inventory balance, cutting pattern yields, shelf life of fresh products, balance among different sections of the animal, warehouse capacities and labour availability. According to previous description the proposed model considers the following notation:

3.1   Sets and Indexes

T : number of periods of the planning horizon.

J : number of cutting patterns.

L : Shelf life for fresh products.

H : Carcasses available to process during the whole planning horizon.

$k \in K$: Set of sections per carcass.

$j \in J_k$: Set of cutting patterns per section k.

$r \in R$: Set to represent the different types of carcasses

$i \in P$: Set of products.

α_r : Proportion of carcasses of type r.

ψ_{ijr} : Yield of product i using cutting pattern j on carcasses of type r.

p_i^f : Selling price per fresh product i .

p_i^c : Selling price per frozen product i.

c_j : Operational cost of pattern j .

c_j^e : Operational cost of pattern j in overtime.

b^c : freezing process cost per kilogram.

h^f : Holding cost of fresh product per period.

h^c : Holding cost of frozen product per period.

s_i^f : Cost for unsatisfied-demand of fresh product i.

s_i^c : Cost for unsatisfied-demand of frozen product i.

F^c : Freezing process capacity.

d_{it}^f : Demand of fresh product i at each period t.

d_{it}^c : Demand of frozen product i at each period t.

τ : freezing process duration.

W^f : Warehouse capacity (in kg.) for fresh products.

W^c : Warehouse capacity (in kg) for frozen products.

t_j : operation time for cutting pattern j.

T_w : Available hours in regular time.

T_w^e : Available hours in overtime.

δ : Auxiliary parameter for better control of the available carcasses.

3.2   Decision Variables

x_{it} : Total quantity of product i to be processed in period t.

$x_{it(t')}^f$: Quantity of fresh product i to process in period t to be sold at t'

z_{jrt} : Number of times to use pattern j on carcass r in period t in regular time.

z_{jrt}^e : Number of times to use pattern j on carcass r in period t in overtime.

x_{it}^c : Quantity of frozen product i to process in t.

v_{it}^f : Quantity of fresh product i to be sold in t.

v_{it}^c : Quantity of frozen product i to be sold in t.

I_{it}^c : Quantity of frozen product i to hold in t.

u_{it}^f : Unsatisfied-demand of fresh product i in t.

u_{it}^c : Unsatisfied-demand of frozen product i in t.

H_t : Number of carcasses to be processed in period t.

3.3   Objective Function

The objective function is obtained by calculating the difference between the incomes from selling the products yielded by the cutting patterns, minus the operational costs incurred. These operational costs involve freezing costs, inventory cost of frozen and fresh products, unsatisfied-demand penalties and labour costs to perform the cutting-patterns, respectively:

$$\max \sum_{i\in P}\sum_{t=1}^{T}(p_i^f v_{it}^f + p_i^c v_{it}^c) - \sum_{i\in P}\sum_{t=1}^{T} b^c x_{it}^c - \sum_{i\in P}\sum_{t=1}^{T} h^c I_{it}^c \qquad (1)$$

$$- \sum_{i\in P}\sum_{t=1}^{T}\sum_{l=0}^{L}(h^f l x_{it(t+l)}^f) - \sum_{i\in P}\sum_{t=1}^{T}(s_i^f u_{it}^f + s_i^c u_{it}^c) - \sum_{t=1}^{T}\sum_{j\in J}\sum_{r\in R}(c_j z_{jrt} + c_j^e z_{jrt}^e)$$

3.4   Constraints

The optimal solution of the model satisfies a different set of constraints described next.

Carcass to be Processed. Cutting patterns are partitioned per sections in order to reduce the number of different patterns, as explained by [10]. The carcasses are cut in different sections, and for each section a different cutting pattern can be applied. These constraints ensure a balance between cutting patterns and the number of carcasses to be processed at each time period. Equality is forced because it is not possible to leave raw material unprocessed.

$$\alpha_r H_t = \sum_{j\in J_k} \left(z_{jrt} + z_{jrt}^e \right) \quad t = 1, \ldots T, \; \forall k \in K, \; \forall r \in R \qquad (2)$$

Carcass Limit. The number of carcasses to process in each period is a decision variable. However, the model imposes a lower and upper limit according to the animal availability from suppliers and a given percentage δ to allow an extra flexibility in the total number of carcasses to be processed.

$$\delta H \le \sum_{t=1}^{T} H_t \le H. \qquad (3)$$

Cutting Patterns Yield. In the meat industry, different cutting patterns can be applied on the carcass to make different products. A cutting pattern is therefore defined by a combination of a set of products and their respective yields. It is assumed that a specific product can be obtained from different cutting patterns, but not from different sections. Products per sections are considered independent, except some specific products, such as skin, bones, and trimming, but they are not considered in the analysis. The following constraint calculates the total kilograms of each product retrieved from all the cutting patterns applied at each time period.

$$x_{it} = \sum_{r\in R}\sum_{j\in J_k} \psi_{ijr} \left(z_{jrt} + z_{jrt}^e \right) \quad \forall i \in P, \; t = 1, \ldots, T. \qquad (4)$$

Available Work Hours per Period. These constraints ensure that the labour time does not exceed the viable working hours in regular time:

$$\sum_{r \in R} \sum_{j \in J} t_j z_{jrt} \leq T_w \quad \forall t = 1, \ldots T, \tag{5}$$

and during overtime:

$$\sum_{r \in R} \sum_{j \in J} t_j z_{jrt}^e \leq T_w^e \quad \forall t = 1, \ldots, T. \tag{6}$$

Fresh and Frozen Balance. This constraint determines the amount of product to be frozen and the amount to keep fresh to be sold in the next period.

$$x_{it} = \sum_{l=0}^{L} x_{it(t+l)}^f + x_{it}^c \quad \forall i \in P, t = 1, \ldots, T. \tag{7}$$

It is recognized that the meat industry works with perishable products subject to spoilage. In order to extend the life of the product, it undergoes a freezing process. Thereby, a product can be sold in two presentations, fresh and frozen. A product is considered fresh if it is sold within a given number of days after elaboration, according to the value given to set L. On the other hand, frozen products can be kept for almost 2 years. However, the profit of selling frozen products are considerably less than fresh products.

Fresh Product to be Sold. As mentioned, fresh products are not allowed to be kept for more than a given number of days. Constraint (8) calculates the total amount of fresh products that can be sold in a period t, but were produced in previous periods in a given time windows.

$$v_{it}^f = \sum_{l=0}^{L} x_{i(t-l)t}^f \quad \forall i \in P, t = 1, \ldots, T. \tag{8}$$

Frozen Product to be Sold. Fresh products need to stay at least 2 days in the freezing tunnel, to be considered frozen. The following constraint balances the inventory of frozen products for each time period.

$$v_{it}^c = I_{i(t-1)}^c + x_{i(t-\tau)}^C - I_{it}^c \quad \forall i \in P, t = 1, \ldots, T. \tag{9}$$

Demand of Fresh Products. Ensures that the requested level of each fresh product is addressed, allowing the existence of unsatisfied-demand if the raw materials are insufficient.

$$v_{it}^f + u_{it}^f = d_{it}^f \quad \forall i \in P, t = 1, \ldots, T. \tag{10}$$

Demand of Frozen Products. Ensures that the requested level of each frozen product is addressed, allowing also the existence of unsatisfied demand.

$$v_{it}^c + u_{it}^c = d_{it}^c \quad \forall i \in P, t = 1, \ldots, T. \tag{11}$$

Freezing Capacity. Fresh products need to be processed during τ periods in order to become frozen. The following constraint ensures that the capacity of this process is never exceeded, assuming that $x^c_{i(1-\tau)}, \ldots, x^c_{i0}$ are given parameters:

$$\sum_{i \in P} \left(\sum_{t'=t-\tau}^{t} x^c_{it'} \right) \le F^c \quad t = 1, \ldots, T. \tag{12}$$

Fresh Products Warehouse Capacity. Ensures that the capacity for holding fresh products is never exceeded.

$$\sum_{i \in P} \sum_{l=1}^{L} \sum_{t'=t}^{t-l+|L|-1} x^f_{i(t-l)(t')} \le W^f \quad t = 1, \ldots, T. \tag{13}$$

Frozen Products Warehouse Capacity. Ensures that the capacity for holding frozen products is never exceeded.

$$\sum_{i \in P} I^c_{it} \le W^c \quad t = 1, \ldots, T. \tag{14}$$

Decision Variables Nature. H_t, z_{jt} and z^e_{jt} are non-negative integer variables for each $j \in J$ and $t \in T$. The rest of the decision variables are all continuous and non-negative.

4   Computational Results

In this section a case study is presented in order to illustrate the suitability and advantages of the proposed optimization model. Basic parameters (such as prices, costs and warehouse capacities) were created using market information gathered from different pork producers. Several countries use different cutting patterns for producing meat products according to their history and gastronomic culture. The case study considers the cutting patterns used by a given Mexican pork firm that must plan its production over time horizon. First, pork carcasses were split up into 5 sections, and for each section a set of cutting patterns was assigned. In total, the company operates with 17 cutting patterns, and manages 40 pork products.

The case study represents a batch of fattened pigs arriving everyday to the meat packing plant to be slaughtered and later processed as carcasses. It is assumed the available amount of carcasses during the whole horizon is fixed and known. The total amount of carcasses available over time horizon was set in 2825. The yield matrix for a carcass per product, section and cutting pattern was obtained from production lines (in kg). Table 1 presents the yield matrix used for Sect. 3. Since the yield matrix will depend on growing conditions, breeding lines, feeding, management schemes, carcass weight, it may exists large and natural variations among the yield obtained for each processed carcass. However, for study proposes it is assumed that the farm maintains a general breeding program allowing to supply a batch of homogeneous pigs, which carcasses are weighting around 63 kg (head, tongue, trotters, and foot are not considered in the carcass weight). The yield matrix from Table 1 clearly shows some products are obtained from different cutting patterns, for instance the product 19 can be obtained from cutting pattern 9, 11, or 12, but the yield per product obtained from each cutting pattern is different (7.65, 7.80 and 5.40 respectively). Moreover, there are other products that are exclusive for a specific cutting pattern, for instance see cutting pattern 10, products 24 and 25).

Table 1. The yield per product, and cutting pattern given in kg for Sect. 3.

Section 3	Products										
Patterns	18	19	20	21	22	23	24	25	26	27	28
8	11.63	0.00	0.00	0.00	0.00	0.00	0.00	0.00	0.00	0.00	0.00
9	0.00	7.65	1.44	0.29	0.43	1.43	0.00	0.00	0.00	0.00	0.00
10	0.00	0.00	0.00	0.00	0.00	0.00	8.23	3.90	0.00	0.00	0.00
11	0.00	7.80	1.91	0.29	0.00	1.43	0.00	0.00	0.00	0.00	0.00
12	0.00	5.40	0.00	0.00	0.00	0.00	0.00	0.00	2.57	1.97	1.19

Labour capacity is considered as 8 h per day in normal time, and 3 h per day in overtime. To perform each cutting pattern a specific amount of labour time is required. Table 2 presents the set of cutting patterns applied to Sect. 3 and the corresponding labour time required to perform them. It is observed that some cutting patterns require the same amount of labour time (see cutting pattern 11, and 12), while others require half of the time (see cutting pattern 8).

Table 2. The labour time required to perform a cutting pattern for Sec. 3.

Patterns	8	9	10	11	12
Time	12	28	16	24	24

The pork products may be sold fresh or frozen. Once a carcass is processed all the products obtained are considered fresh, and depending on demand some products are directly sent to the freezing process, and others to the inventory for later selling. The shelf life for fresh product is set to $L = 3$ days, while the freezing process takes $\tau = 2$ days. The demand for fresh and frozen products is assumed to be known over time horizon. For instance, Table 3 presents the demand over time horizon for fresh products of Sect. 3. It is observed demand is not homogeneous over time and among products.

The solved instance considered 10 planning periods, this results in a model with 6601 variables from which 1750 are integer, and 3155 constraints.

Table 3. The demand over time horizon for fresh products of Sect. 3.

Product	Horizon period									
	1	2	3	4	5	6	7	8	9	10
18	799	467	677	3453	345	765	452	658	1324	234
19	346	5743	7657	3452	6256	3562	253	3567	355	243
20	36	63	678	442	564	341	556	34	155	254
21	43	11	23	53	85	24	78	90	34	63
22	735	89	45	244	534	256	789	323	24	4234
23	645	0	95	345	657	35	645	3646	67	465
24	256	4641	5731	0	7869	66	2346	446	44	674
25	345	673	434	4545	246	234	665	785	764	345
26	566	5536	567	345	123	546	0	0	546	123
27	234	212	0	435	24	131	256	123	233	231
28	245	345	234	5356	345	2345	3452	445	245	525

The results from the case study shows a net profit of $321802.85 dollars, sales per $402288.81 dollars and operative cost per $80485.96 dollars. In order to see the advantages of the model, the results are compared with the decision making of the company see Table 4.

Table 4. The number of time each cutting pattern is applied over time horizon.

	Model					Company				
Section 1	**P1**	**P2**	**P3**	**P4**		**P1**	**P2**	**P3**	**P4**	
	269	878	509	1169		269	979	642	935	
Section 2	**P5**	**P6**	**P7**			**P5**	**P6**	**P7**		
	1085	1740	0			839	1556	430		
Section 3	**P8**	**P9**	**P10**	**P11**	**P12**	**P8**	**P9**	**P10**	**P11**	**P12**
	788	0	1909	128	0	788	215	1194	278	350
Section 4	**P13**	**P14**	**P15**			**P13**	**P14**	**P15**		
	2000	825	0			1650	915	260		
Section 5	**P16**	**P17**				**P16**	**P17**			
	1920	905				1990	835			

The model gets 8 % more profit than the one obtained by the company. Also, it can be notice that the model takes a better advantage of the resources than the company. This is seen in the Table 5. The amount of inventory for frozen products is bigger for the company, because fresh products obtained do not meet the demand, and need to be frozen. Furthermore, the percentage of use of labour

Table 5. Percentage of use of warehouse capacity for frozen products.

	Period									
Output	1	2	3	4	5	6	7	8	9	10
Model	0	11	18	23	36	45	48	50	41	41
Company	0	23	37	45	71	90	95	98	81	80

capacity in normal time is higher for the model (80 %) than for the company (77 %), while the percentage of use of over time is higher for the company (31 %), than the model (23 %). The results obtained are directly related with the way and time the manager decide how to cut up the carcasses.

All instances described in the computational results are implemented and solved using a PC with Windows 7, i5-3210M CPU @ 2.5 GHz and 8 Gb RAM, and the optimization software package IBM ILOG CPLEX Optimization Studio 12.6. The computational time required to solve the problem in all instances where less than 2 min.

5 Conclusions and Future Research

In this paper a mixed integer linear programming model for production planning in the meat industry is proposed and validated. This model considers aspect such as cutting patterns per sections, multi-products, finite planning horizon, product perishability, freezing and cutting operation costs, normal and over time, and carcasses availability. Major decisions imply the number of times each cutting pattern is applied on the carcasses. But, the amount of carcasses to be processed act as a decision variable with both lower and upper bounds. The lower bound represents the minimal amount of carcasses to be processed to cover operative cost, while upper bound the amount of available carcasses, considering in this way the pig supplier (fattening farms)into the model.

The results explained in the previous sections show the formulated model is highly applicable due to low resolution times (all instances are solved within a minute and less than 0.05 % gap) and gives high-value information for the decision makers. However, for large instances, by example the ones generated with horizon periods longer than a month, and considering more products and cutting patterns; the model substantially grows of size and thereby it can not be solvable for the exact algorithm of Branch and bound.

Future research in this area should go towards the development of large and complex instances that better represent the planning operation of a meat packing plant, and the design, development and analysis of efficient algorithms to solve such problems.

Acknowledgements. This research was partially supported by CONICYT, Departamento de Relaciones Internacionales "Programa de Cooperacion Cientifica International" (Grant PPCI 12041), DGIP of Universidad Tecnica Federico Santa Maria

(Grant USM 28.15.20), and by Consejo Nacional de Ciencia y Tecnologa, CONACYT, Grant (No. 188902).

References

1. Ahumada, O., Villalobos, J.: Application of planning models in the agri-food supply chain: a review. Eur. J. Oper. Res. **196**(1), 1–20 (2009)
2. Bixby, A., Downs, B., Self, M.: A scheduling and capable-to-promise application for swift & company. Interfaces **36**(1), 69–86 (2006)
3. Bjorndal, T., Herrero, I., Newman, A., Romero, C., Weintraub, A.: Operations research in the natural resource industry. Int. Trans. Oper. Res. **19**, 39–62 (2012)
4. Higgins, A., Miller, C., Archer, A., Ton, T., Fletcher, C., McAllister, R.: Challenges of operation research practice in agricultural value chains. J. Oper. Res. Soc. **61**(6), 964–973 (2010)
5. Pla, L., Sandars, D., Higgins, A.: A perspective on operational research prospects for agriculture. J. Oper. Res. Soc. **65**, 1078–1089 (2014)
6. Reynisdottir, K.: Linear optimization model that maximizes the value of pork products. Master's thesis, Reykjavik University (2012)
7. Sanchez, R. A.: Modelo para Optimizar la planificacion de la Produccion de Productos y Subproductos en la Industria Porcina. Master's thesis, Universidad de Talca, Chile (2011)
8. Sanabria, L.: Un Modelo de Planeacion de la Produccion para una Planta Empacadora de Carnicos. Master's thesis, Universidad Autnoma de Nuevo Leon, Mexico (2012)
9. Stokes, J.R., Sturdivant, A.W., Rister, M.E., Mccarl, B.A.: Meat packing plant production planning: application of mixed integer goal programming. Agribusiness **14**(3), 171–181 (1998)
10. Whitaker, D., Cammel, S.: A partitioned cutting-stock problem applied in the meat industry. J. Oper. Res. Soc. **41**(9), 801–807 (1990)
11. Wikborg, U.: Online meat cutting optimisation. Master's thesis, Norwegian University of Science and Technology, Norway (2008)

The Non-Emergency Patient Transport Modelled as a Team Orienteering Problem

José A. Oliveira[⊠], João Ferreira, Luis Dias, Manuel Figueiredo,
and Guilherme Pereira

ALGORITMI Research Centre, University of Minho, Guimarães, Portugal
Jose.Oliveira@algoritmi.uminho.pt,
joao.aoferreira@gmail.com,
{lsd,mcf,gui}@dps.uminho.pt

Abstract. This work presents an improved model to solve the non-emergency patients transport (NEPT) service issues given the new rules recently established in Portugal. The model follows the same principle of the Team Orienteering Problem by selecting the patients to be included in the routes attending the maximum reduction in costs when compared with individual transportation. This model establishes the best sets of patients to be transported together. The model was implemented in AMPL and a compact formulation was solved using NEOS Server. A heuristic procedure based on iteratively solving Orienteering Problems is presented, and this heuristic provides good results in terms of accuracy and computation time. Euclidean instances as well as asymmetric real data gathered from Google maps were used, and the model has a promising performance mainly with asymmetric cost matrices.

Keywords: Non-emergency patients transport · Team Orienteering Problem · Mixed Integer Linear Programming · AMPL · NEOS Server

1 Introduction

In 2012, Portuguese Authorities published several official documents concerning the Non-Emergency Patient Transport (NEPT) service in "Diário da República" (DRE). The legislative motivation was based on a requirement laid down in the Memorandum of Understanding signed between the Portuguese Government and the International Monetary Fund, the European Central Bank and the European Union to reduce the cost of NEPT services [13]. This legislation mandates the minimum requirements for the quality and safety of care delivered to patients by NEPT services, [14].

In 2011, the Northern Department of Health [3] reported the implementation of a computer system to optimize the management of NEPT services. The published documents state that the computer system "will allow greater accuracy in terms of prescription; will ensure the organization in a rational way; will promote the multiple patients transport." The Northern Department of Health expected to achieve a reduction of €3 million on transport costs in the region, representing a 20 % cost reduction. Oliveira et al., [29] showed that it was possible to obtain a significant reduction in the costs in their experiments. They used an euclidean cost matrix in their instances and

© Springer International Publishing Switzerland 2015
D. de Werra et al. (Eds.): ICORES 2015, CCIS 577, pp. 147–164, 2015.
DOI: 10.1007/978-3-319-27680-9_10

they achieved a reduction between 4 % and 6.5 % in average per each vehicle considered.

This work presents an improved model to solve the non-emergency patients transport (NEPT) service issues given the new rules mentioned above - following the same principle of the Team Orienteering Problem, selecting the patients to be included in the routes attending the maximum reduction in costs and establishing the best sets of patients to be transported together.

The current paper is composed of five sections. After the introduction, Sect. 2 presents a description of the problem, the main topics referred by the law, and a brief literature review. Section 3 describes the model and presents the mathematical formulation. Section 4 includes a discussion of the heuristic and its results. Section 5 summarizes the main conclusions of this work.

2   Transport Problem

2.1   Nept Definition

In light of the legislation in 2014 in Portugal [15–21], and with regard to access by the users to the services of the National Health Service (NHS), NEPT is considered the associated transport system for the health care system, where the origin or destination are the medical centres and services within the NHS, private entities or social entities with a contract or agreement for the provision of health care under the following conditions: consultations for inpatient or outpatient surgery, diagnostic procedures and therapeutic treatments, transporting the patient after discharge from hospital (with prior prescription) and transporting the patient after discharge from the emergency room (with prior prescription).

2.2   Transport Prescription

Prescription of transportation is solely the responsibility of the attending physician, who shall record the following information in the support doctor system or equivalent system: the clinical justification, or reasons for needing transport, and verification of the economic condition. Where there is a need to make the ambulance transport, the following is required: the justification of the mode of transport; the conditions under which the transport should occur, particularly if the patient requires ventilation, oxygen, wheelchair or is sick in bed or isolated, the justification of the need for a companion, and the justification of the need for monitoring a health professional. After the shipping prescription by the physician is completed, a member of the administrative services staff validates the economic condition of the patient and proceeds to request transportation.

The requisition of transport satisfies the criteria of minimizing the distance between the place of origin, which must match the address from which the transportation is effected, and the place of destination, which must take into account the location of the nearest place of origin. The NEPT is performed by ambulance or by ambulette (simple vehicle for patient transport - SVPT), which is a passenger car, with a maximum capacity of five or nine people, for the NEPT service whose clinical condition does not impose the need for health care during the transport.

The non-emergency patient transport is carried out, where possible, in SVPT, taking into account the need to optimize the capacity of the vehicle against the following criteria: (a) Grouping of clients, regardless of origin, within the same route; (b) intended for health facility and preferably in the same county or geographical area; (c) users for the same time period for consultation or treatment.

The attending physician shall justify the use of individual transport ambulances, in a reasoned manner. To further the grouping of users, the NEPT may allow deviations of less than 10 km or 30 min journey, considering the travelling of the first patient. The first patient assumes a critical importance to define the route and to define the cost of the transportation service. The first patient should be the most distant patient to destination. A transport on SVPT may carry a single patient in cases where there are no other patients in the same time period or along the same route, but this is an exception, and an effort must be made to carry additional patients where possible.

2.3   Transport Organization

Transportation must be ordered and scheduled at least 48 h before the journey. In individual situations of an exceptional nature where this time limit cannot be observed since no agreement has been authorized between the requesting entity and the carrier, the time limit of tacit acceptance and approval of daily transportation requests is determined by the computer system, at 15 h and 30 min.

Requests for transportation are grouped according to the schedules of supply of care to which patients are intended, according to the following guidelines:

- If the destination is within the geographic boundaries of the patient's county of origin or within a radius of 10 km, these requests are grouped at one hour intervals between delivery of the first and the last patient;
- If the destination is outside the patient's county of origin, the interval between the delivery of the first patient and the last can be two or four hours, depending on the distance which, according to geographical features, can be a range of 100 km to 130 km. This time is to be determined by the entity responsible for organizing the transport.

The law states that the rule of the maximum deviation should be applied to the "Optimization of routes". Patients can be grouped for a journey, regardless of parish or municipality, provided that there is no deviation more than 10 km or 30 min, provided the previous guidelines are observed.

The costs resulting from NEPT are the responsibility of the requesting entity. Thus it becomes important for the requesting entity to optimize and streamline the process of transporting non-emergency patients. It is not known how the requesting entity must make the optimization of NEPT, so this study hopes to be an important and timely contribution.

2.4   Literature Review

The health care industry is rife with problems of management and organization which have been studied over the past several decades [32]; [4]. The research attention to such problems is increasing and, in the Western world, results from increased demand for health care and the need to keep the social cost of health care as low as possible. The increased demand for health care has two vectors justifying its growth: the democratization of access to healthcare in developed societies and an aging population. The increased demand for health care causes transporting patients an important problem in the logistics of the health systems, since it is a significant portion of operating costs. As far as health logistic problems are concerned, an effective transport service is now becoming an extremely complex problem that has to be solved efficiently, so it requires the best solution methodologies. Bellamy et al. [5] defines non-emergency transport needs including ordinary situations in which a patient simply cannot get to and from a healthcare facility under their own power.

According to Wilken et al., [35], patient transportation is a critical part in providing healthcare services. The authors discuss non-emergency transportation in rural southern Illinois, and they point out the importance of this issue. Many individuals do not have the funds to pay for public transportation and often public transportation is not available or easily accessible to them so they may miss regular doctor's appointments, treatments, and so on. As a consequence, the patient may become more ill and then must be transported to a medical facility by emergency transportation. Safaei [31] studied the non-emergency medical transportation services available in rural British Columbia, and he reported the success of the "Connections service" in enhancing rural and northern communities' access to healthcare services, in particular among those with limited means and resources.

Health authorities and health managers rely on knowledge and state of the art technology to address the logistic of health systems. Today, information systems provide support for making the decision, allowing gains in effectiveness and efficiency. Transport in health care is a subject which has been studied in the literature of optimization of logistic processes for many years. Vaisblat and Albert [33] carried out one study which focused on the scheduling of a special fleet serving the needs of patients. Hains et al., [25] addressed the issue of safety and quality in NEPT. Recently Díaz-Parra et al., [11] published a state of the art review on the problem of transport which included several variants of problems, mathematical formulas, and techniques used in problem solving.

One of the most studied routing problems is the vehicle routing problem (VRP), which basically aims to define a set of vehicle routes which are optimized to visit a series of well-defined locations. This problem presents a large number of variants which address more specific situations. Berbeglia et al., [6] refer to the static or dynamic nature of routing problems. In the static case, all the information is known in advance and does not change during the construction and implementation phases. In the dynamic case, the available information is updated (changed) during both the construction phase and the implementation phase, by virtue of new user requests. In these problems, the proposed solution is a solution strategy that can be changed with the passage of time. Typically, dynamic situations of this kind occur in transport at the request of users with special needs, which need to be sent to the car which will transport them to their destination. The dynamic aspect of this problem stems from the fact that transportation requests sometimes arise on the same day they need to be met: this type of problem is called a DARP (Dial-A-Ride Problem).

Both the static and dynamic versions of DARP have received significant contributions, such as a review of Cordeau and Laporte [9] about models and algorithms. Psaraftis [30] examined a single route of this problem in which clients request a service to be made available as soon as possible. Whenever a new order is entered, the system updates the proceedings and attempts to accommodate the new request on an existing, but only partially complete, route. Meanwhile, Madsen et al. [28] presented an algorithm to a real case of the dynamic DARP with multiple vehicles that met up to 300 requests daily to transport people with special needs.

One VRP variant is the "Vehicle routing problems with profits" [2]. In this type of problem, two different decisions have to be made simultaneously—which which customers to serve and how to sequence them in one or several routes. In general, a profit is associated with each customer that makes the customer more or less attractive. The majority of real-world applications require systems that are more flexible in order to overcome some imposed constraints that may lead to the selection of customers. To deal with the selection of customers, the Team Orienteering Problem (TOP) models can be used. The main difference between the TOP and the VRP is related to the fact that not all the TOP vertices of the graph (clients) must be visited, as in the VRP. In the TOP, each customer has an associated profit, and the routes have maximum durations or distances. The choice of customers is made by balancing their profits and their contributions to the route duration or distance. The objective is to maximize the total reward collected by all routes while satisfying the time limit.

The TOP is a fairly recent concept, first suggested by [7] under the name Multiple Tour Maximum Collection Problem. Later, Chao et al., [8] formally introduced the problem and designed one of the most frequently used sets of benchmark instances. TOP has received significant attention from the scientific community [34]; [2] either in presenting exact solution methodology [1] or in approximate solution methodology [26]. Vansteenwegen and his team [34] maintain a repository of public instances (The Orienteering Problem: Test Instances, 2014).

Gutiérrez-Jarpa et al., [24] studied the problem with fixed delivery and optional collections, utilizing a mixed solution which uses VRP for delivery and TOP for collection. The authors studied the particular case of a single vehicle and presented a new branch-and-cut method that allows the system to solve larger instances. The method can solve instances which include up to 90 vertices. The authors refer to the need to extend the investigation to cases with multiple vehicles and the development of heuristics to solve large scale instances. Despite the great practical interest that this modelling system has for reverse logistics, the authors report they found only study, Gribkovskaia et al., [22] that have applied tabu search to the single vehicle pickup and delivery problem with selective pickups.

3   The Model

The real problem studied in this article concerns the non-emergency transport of patients from their homes to the hospital and from the hospital back to their homes. Currently, in Portugal the shuttle typically collects patients from their homes to the hospital for treatment and back again. This system creates the situation – "many

152   J.A. Oliveira et al.

(origins)-to-one (target)-to-many (destinations)." The way to organize this transport is not clearly established in law, which means that money is being wasted on the waiting time of the vehicle, and patients often waste time waiting for their transportation.

We decided by modelling the NEPT problem with the TOP point of view. What will happen is that given a list of non-urgent patients for whom transportation was requested and given a fleet of vehicles available with a capacity of eight seats, it is the allocation of transport services to maximize the occupancy of the vehicle and minimizing the distance travelled. Patients who can not be included in the routes of these vehicles will be transported in specially requested for this ambulances service.

3.1   Mathematical Model

Since the group of patients is a severely constrained situation, it is our choice to model the real problem presented by the NHS as a Team Orienteering Problem, solving for the set of available vehicles. Vertices not included in the routes of the problem are the users who will make the path by individual transport.

Our model is based on the Team Orienteering Problem and we follow the mathematical model presented by Labadie et al., [27]. We define the search of different paths from a common start point ($i = 1$) to a common ending point ($i = n$). If we want the start point to be the same ending point, we use the same coordinates for both points.

We have established the following variables:

x_{ij}^k - set of binary variables that is equal to 1 if arc (i, j) is selected in the path k and 0 otherwise.

y_j^k - set of binary variables that is equal to 1 if vertex j is in path k and 0 otherwise.

w_j^k - set of binary variables that is equal to 1 if vertex j is the first vertex in the path after the start point.

W^k - is the maximum value for the length of the path that is a function of the distance of the first vertex to terminal vertex and an allowed increase.

X_j - a variable the controls a sequential number for the vertices in the path.

TPS_{ij}^k - is an auxiliary variable to linearize the product of $w_i^k * y_j^k$.

To define the objective function we define three parcels. TIC is the Total Individual Cost, equivalent to transporting each patient individually. CIP This parcel is the main savings when a patient is not transported individually. APC is Additional Patient Cost (incremental cost) related to patients who are transported together in same vehicle with the first patient. This parcel is to pay the additional deviations to collect patients in the route of the first patient. By law, this cost is nowadays 20 % of the cost of the first patient in the route. In a solution with an individual transport for all patients, CIP and APC are equal to zero.

$$TIC = \sum_{i=2}^{n-1} c_{1i} + c_{in}$$

$$CIP = \sum_{k}^{m} \sum_{j=2}^{n-1} \sum_{i=2}^{n-1} (c_{1,j} + c_{jn}) x_{ij}^{k}$$

$$APC = \sum_{k=1}^{m} \sum_{j=2}^{n} \sum_{i=1}^{n-1} s(c_{1i} + c_{in}) TPS_{ij}^{k}$$

Figure 1 explains these calculations considering the transportation of two patients: A and B. Figure 1(a) COST1 is the solution cost using two vehicles/routes, both starting at S and ending at E; Fig. 1(b) COST2 is the solution cost using only one vehicle, where A is first patient in the route; while Fig. 1(c) COST3 is the solution cost using one vehicle, where B is the first patient in the route.

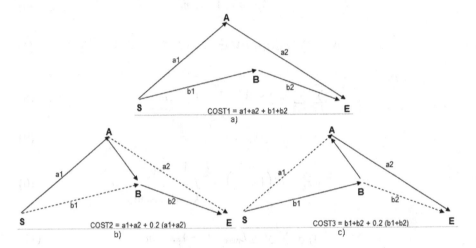

Fig. 1. Cost parcels.

It is possible to establish the following relations:
COST1 = TIC;
CIP(A) = a1 + a2; CIP(B) = b1 + b2;
COST2 = TIC-CIP(B) + APC(B);
COST2 = TIC-CIP(B) + 20 %CIP(A);
COST3 = TIC-CIP(A) + APC(A);
COST3 = TIC-CIP(B) + 20 %CIP(B).
Obviously, APC(A) and APC(B) depends on which is the first patient in the route where they are included.

The mathematical formulation of the Mixed Integer Linear Programming (MILP) is presented next:

$$\min(TIC - CIP + APC) \qquad (1)$$

subject to:

$$\sum_{i=1}^{n-1} x_{ij}^k = \sum_{i=2}^{n} x_{ji}^k \quad j = 1, \ldots, n-1; k = 1, \ldots, m \tag{2}$$

$$\sum_{i=1}^{n-1} x_{ij}^k = \sum_{i=2}^{n} x_{ji}^k \quad j = 2, \ldots, n; k = 1, \ldots, m \tag{3}$$

$$\sum_{j=2}^{n} x_{1j}^k = \sum_{i=1}^{n-1} x_{in}^k = 1 \quad k = 1, \ldots, m \tag{4}$$

$$\sum_{k=1}^{m} \sum_{i=1}^{n-1} x_{ij}^k \leq 1 \quad j = 2, \ldots, n-1 \tag{5}$$

$$w_j^k = x_{1j}^k \quad k = 1, \ldots, m \tag{6}$$

$$W^k = \sum_{j=2}^{n-1} c_{jn} x_{1j}^k + dMax \quad k = 1, \ldots, m \tag{7}$$

$$\sum_{i=2}^{n-1} \sum_{j=2}^{n} c_{ij} x_{ij}^k \leq W^k \quad k = 1, \ldots, m \tag{8}$$

$$X_1 = 1 \tag{9}$$

$$X_j \geq X_i + x_{ij}^k - M\left(1 - x_{ij}^k\right) \quad \begin{array}{l} i = 1, \ldots, n-1; \\ j = 2, \ldots, n; \\ k = 1, \ldots, m \end{array} \tag{10}$$

$$\sum_{i=2}^{n-1} w_i^k + \sum_{i=2}^{n-1} y_i^k \leq L_{\max} \quad k = 1, \ldots, m \tag{11}$$

$$\begin{array}{l} TPS_{ij}^k \leq w_i^k \\ TPS_{ij}^k \leq y_j^k \\ TPS_{ij}^k \geq w_i^k + y_j^k - 1 \end{array} \quad \left\{ \begin{array}{l} i = 2, \ldots, n-1 \\ j = 2, \ldots, n-1 \\ k = 1, \ldots, m \end{array} \right. \tag{12}$$

Expression (1) represents the objective function to be minimized. It is intended to diminish the total transport cost removing individual transportation as much as possible, paying the necessary deviations to collect patients in shared routes. Once TIC is a constant value, the objective function could be rewritten in the following way:

$$\max(CIP - APC) \tag{13}$$

The expression (13) is more close to TOP definition, where the *profit* represents the saving relatively the individual transportation.

In terms of constraints, expression (2) assign visited patients to only one route, and in expression (5) patients could be visited by only one route. Expression (3) ensures the flow conservation in each node. Expression (4) ensures that a vehicle starts the route from node 1. Expressions (6) (7) and (8) establish the first patient in the route and calculate the maximum length for the route according to the distance from the first patient to the destination. The law allows an increase in the length of the route to collect additional patients, but this is currently limited to 10 km or 30 min [15].

Expressions (9) and (10) eliminate sub tours, and the capacity of vehicle is verified in expression (11). Expression (12) linearizes the objective function.

3.2   NEOS Server Experiments

The model was implemented in AMPL language and submitted to the NEOS Server to evaluate the quality of solutions provided by this compact formulation [10, 12, 23]. Sixty-four Euclidean instances were randomly created to perform the computational experiments. The instances are divided into two sets according to the capacity of the vehicle four places or eight places available to transport the patients. The size of the instances varies from twenty to one hundred patients, and from three to ten vehicles.

Euclidean Instances. Using the NEOS Server with AMPL/Gurobi/MINTO/scip/ XpressMP, the MILP could not find a solution for all instances. Experiments with instances of different sizes were performed to find the maximum number of vertices that it is possible to solve optimally. Memory errors ("mem_error") were reported when the solution exceed 3 GB of memory limit. Also, it is only possible to use a maximum of eight hours of computation with the NEOS Server. When the maximum time was achieved, it reported the best solution founded and the correspondent gap. Tables 1–2 present these results.

Apparently solving instances with vehicles with larger capacity becomes easier and it was possible to solve an instance with one hundred patients and three vehicles.

Asymmetric Real Data Instances. In order to evaluate the capacity of treating real data, several instances of similar size to previous ones were generated. Figure 2 shows

Table 1. Experimental results (capacity = 4).

vehicles x capacity

nodes	3x4	4x4	7x4	10x4
20	1272.65	1184.0	1055.93	1055.93
30	1840.37	1693.42	1383.03 gap 5.43%	1299.10 gap 34.16%
40	2671.87	2498.09	2048.70 gap 20.73%	1776.02 gap 70.69%
50	3478.46	3248.65	2709.00 gap 17.63%	2362.49 gap 47.45%
60	4311.22	4074.27 gap 3.42%	3455.71 gap 17.67%	2987.28 gap 36.01%
70	5043.11	4802.52 gap 3.29%	4131.28 gap 12.73%	3718.51 gap 31.35%
80	5818.70	--- mem_error	4931.43 gap 12.31%	4507.76 gap 30.52%
100	---- mem_error	7155.63 gap 4.55%	--- mem_error	8198.99 gap 77.61%

Table 2. Experimental results (capacity = 8).

nodes	vehicles x capacity			
	3x8	4x8	7x8	10x8
20	1245.22	1156.65	1048.91	1048.91
30	1812.94	1665.99	1376.02	1299.10 gap 30.13%
40	2561.95	2362.82	1915.61 gap 23.26%	1706.82 gap 66.38%
50	3238.03	2974.04	2362.52 gap 31.87%	2075.69 gap 68.38%
60	4003.97	3702.97	2932.47 gap 34.77%	2523.03 gap 83.67%
70	4693.05	4389.54 gap 6.88%	3682.41 gap 36.55%	3135.39 gap 81.95%
80	5450.52	---- mem_error	4332.84 gap 32.28%	3686.95 gap 70.91%
100	6785.59	6505.46 gap 13.98%	---- mem_error	---- mem_error

the dispersion of 100 patients from the county of Guimarães, in the north of Portugal. This instance represents a mix of urban and rural areas.

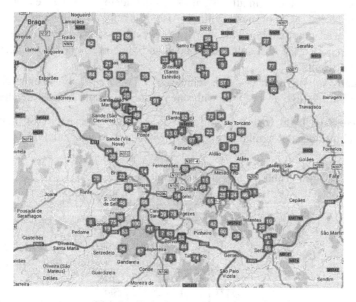

Fig. 2. An instance with real data.

Table 3 presents the objective function value. The values are meters based. Once the real situation does not have so many alternatives as in the euclidean instances, NEOS Server obtain the optimal solutions for all instances with three vehicles with

Table 3. Experimental results (capacity = 4).

vehicles x capacity

nodes	3x4	7x4	10x4
20	247023.2	196907.2 gap 17.5%	196907.2 gap 24.9%
40	582467.0	443055.8 gap 27.7%	368941.0 gap 55.1%
60	991850.6	810441.0 gap 18.7%	781293.2 gap 39.1%
80	1365844.6	1212470.4 gap 15.7%	1207974.0 gap 32.0%
100	1737280.6	1564430.6 gap 12.4%	1729962.2 gap 40.9%

capacity for 4. For the remaining instances it was not possible to obtain the optimal solution within 8 h of CPU, with 4 threads in the NEOS Server.

Table 4 present 15 instances of real data with vehicles with capacity for 8. The optimal value was obtained only for 4 instances with 3 vehicles.

Table 4. Experimental results (capacity = 8).

vehicles x capacity

nodes	3x8	7x8	10x8
20	226450.6	196309.8 gap 19.4%	196309.8 gap 22.4%
40	489604.6	354653.8 gap 48.2%	372290.0 gap 54.6%
60	847047.6	668289.8 gap 50.8%	604410.2 gap 60.9%
80	1205382.4	939195.2 gap 39.1%	831417.6 gap 61.3%
100	1546465.4 gap 1.05%	1357172.2 gap 36.7%	1040835.6 gap 49.2%

NEOS Server did not report memory failures with these real data instances. In general NEOS Server did not found the optimal value within 8 h of computation.

4 Heuristic Procedure

As expected, the compact model could not solve large instances using the NEOS Server. However, the NEOS Server can solve the large instances in terms of vertices using only one vehicle (Orienteering Problem - OP). Given this situation, we developed a heuristic procedure to solve the TOP that is based in successive OP solutions.

Iteratively, for the remaining unvisited vertices, we solve the problem using the compact formulation with one vehicle.

Euclidean Instances. Figure 3 presents the solution obtained with a heuristic procedure to solve the largest instance (one hundred patients with ten vehicles of eight patients capacity). The patients not included in these ten routes must be transported individually. The computational time required to produce the 10 routes was 198 s.

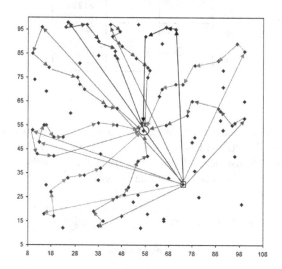

Fig. 3. Heuristic Solution (100 nodes_10 × 8).

To confirm the results obtained with this heuristic, we solved the 32 instances with capacity equal to 8. Table 5 compares the results obtained with NEOS Server ("NeosS" line) and heuristic ("heur" line) in this set of instances.

The heuristic obtained the optimal solution for 13 instances, and obtained better or equal result than NEOS Server in 23 of 32 instances, representing around 70 %. We remind that for some instances we present the values obtained by the NEOS Server at the end of available computation time. These instances are marked with an asterisk "*".

Asymmetric Real Data Instances. Table 6 presents the same type of comparison as previous one, considering now real data instances for vehicles with a capacity of 4 patients.

In Table 7 the heuristic solutions for the real data instances with largest vehicles are presented and compared with the values obtained by NEOS Server.

Again, for some instances (marked with "*"), we present the values obtained by the NEOS Server at the end of available computation time (8 h).

In general, the heuristic produce good results and we point out the optimal values obtained in the shortest instances in terms of available fleet, mainly in the case of vehicles with lowest capacity (Table 6). In all the cases (Tables 5, 6 and 7), the heuristic produces a lower GAP in a very short CPU time, comparing with NEOS

Table 5. Experimental results (capacity = 8).

vehicles x capacity

nodes		3x8	4x8	7x8	10x8
20	NeosS	1245.22	1156.65	1048.91	1048.91
	heur	1245.22	1156.65	1048.91	1048.91
30	NeosS	1812.94	1665.99	1376.02	1299.10*
	heur	1812.94	1665.99	1376.02	1311,43
40	NeosS	2561.95	2362.82	1915.61*	1706.82*
	heur	2561.95	2362.82	1930,27	1748,65
50	NeosS	3238.03	2974.04	2362.52*	2075.69*
	heur	3240.36	2994.45	2382.93	2118,02
60	NeosS	4003.97	3702.97	2932.47*	2523.03*
	heur	4003.97	3707.29	2917.65	2478.83
70	NeosS	4693.05	4389.54*	3682.41*	3135.39*
	heur	4693.05	4389.54	3624.17	3163.34
80	NeosS	5450.52	----	4332.84*	3686.95*
	heur	5450.52	5141.44	4240,82	3627,75
100	NeosS	6785.59	6505.46*	----	----
	heur	6785.59	6403.38	5398.40	4801.58

Table 6. Experimental results (capacity = 4).

vehicles x capacity

nodes		3x4	7x4	10x4
20	NeosS	247023.2	196907.2*	196907.2*
	heur	247023.2	208905.2	208905.2
40	NeosS	582467.0	443055.8*	368941.0*
	heur	582467.0	421172.0	335127.2
60	NeosS	991850.6	810441.0*	781293.2*
	heur	991850.6	790533.2	679204.2
80	NeosS	1365844.6	1212470.4*	1207974.0*
	heur	1365844.6	1157539.6	1017754.4
100	NeosS	1737280.6	1564430.6*	1729962.2*
	heur	1737280.6	1518002.4	1368012.0

Server, in particular in the largest instances. When the comparison with optimal values (Column 1 in Table 7) is possible, the heuristic produces a deviation inferior to 0.45 %.

Figure 4 shows a route obtained with real data to transport eight patients in the same vehicle. The route has 34.4 km length and Google Maps estimates around one hour of travel time.

To collect three of the patients in the route of Fig. 4 the sequence C–B–A is used (Figure 5). This sequence respects constraint (7), because the patient C is more distant (to the destination) than patients B and A, so he must be collected first.

This situation confuses the transport provider, because he needs to go back on the path to where he had passed previously. The sequence A–B–C illustrated in Fig. 6 makes more sense to him, and the total route length is shorter 1 km. However, this sequence does not respect the law that is established in Eq. (7).

This apparent anomaly in the model for generating routes will be reviewed and discussed in future work. The findings of that study may contribute for specific

Table 7. Experimental results (capacity = 8).

vehicles x capacity

nodes		3x8	7x8	10x8
20	NeosS	226450.6	196309.8*	196309.8*
	heur	**226450.6**	210864.0	210864.0
40	NeosS	489604.6	354653.8*	372290.0*
	heur	490233.0	355197.4	**355197.4**
60	NeosS	847047.6	668289.8*	604410.2*
	heur	860820.4	**619896.2**	**534220.0**
80	NeosS	1205382.4	939195.2*	831417.6*
	heur	1205692.0	**882948.6**	**729461.0**
100	NeosS	1546465.4*	1357172.2*	1040835.6*
	heur	**1546217.4**	**1164589.0**	**944620.0**

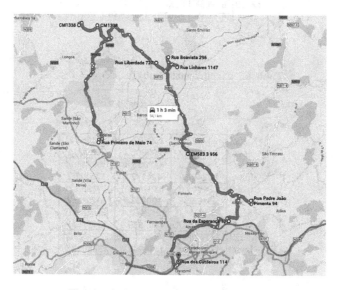

Fig. 4. A route to collect several patients.

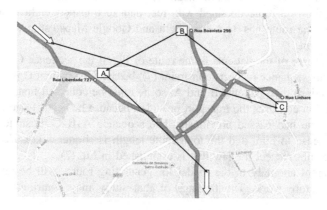

Fig. 5. A detail of the route.

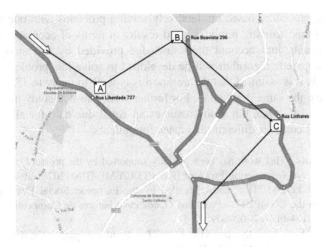

Fig. 6. A sequence with shortest distance.

suggested changes to the law to enable the construction of more environmentally friendly routes, eliminating some unnecessary sub-routes.

5 Conclusions

This work presents a model and a heuristic to solve the problems posed by the non-emergency patient transport in Portugal, given the new rules recently established. The model follows the same principle of the Team Orienteering Problem to select the patients and the routes to be included, providing the maximum cost reduction. This approach is different from VRP strategies once some vertices are not visited. In fact, a patient not visited by the routes would represent a patient that must be transported individually. Indeed, this model establishes the best sets of patients that should be transported jointly.

In this study, several instances (Euclidean and asymmetric real data) were generated to test our approach. The model was implemented in AMPL and our compact formulation was used to solve the instances using the NEOS Server. Instances with one hundred patients and ten vehicles with a capacity for eight patients each could not be solved within available computation time provided by the NEOS Server.

Overall, the computational experiments show that on average a vehicle with 4 seats allows a cost reduction of about 5 % while an 8 seats vehicle permits a reduction of approximately 7 %, when considering a fleet up to 10 vehicles. Computational experiments show that a 10-vehicle fleet reduces up to 55 % of the costs using a shared transport model that considers a common destiny for all patients, compared to individual transportation.

Finally, our model was tested with real instances with distances provide by Google Maps and generates promising preliminary results.

A heuristic procedure based on iteratively solving problems with one vehicle was presented, and this heuristic provides good results in terms of accuracy and computation time. Taking into account the knowledge provided by this study, a greedy heuristic and a genetic algorithm will be developed to solve this problem.

In this work, it is assumed that the transport is type "1 to many to 1", meaning that all patients have the same destination. For further work, we will study the situation of several destinations. Also, this study assumes an equal due date for all patients, but future work will consider different due dates for patients.

Acknowledgements. This work has been partially supported by the project GATOP - Genetic Algorithms for Team Orienteering Problem (Ref PTDC/EME-GIN/120761/2010), financed by national funds by FCT/MCTES, and co-funded by the European Social Development Fund (FEDER) through the COMPETE - Programa Operacional Fatores de Competitividade (POFC) Ref FCOMP-01-0124-FEDER-020609.

This work has been partially supported by FCT – Fundação para a Ciência e Tecnologia within the Project Scope: UID/CEC/00319/2013.

The authors would like to thank the NEOS Server at the University of Wisconsin in Madison for providing support to this study.

References

1. Archetti, C., Bianchessi, N., Speranza, M.G.: Optimal solutions for routing problems with profits. Discrete Appl. Math. **161**(4), 547–557 (2013)
2. Archetti, C., Speranza, M.G., Vigo, D.: Vehicle routing problems with profits. Technical Report WPDEM2013/3, University of Brescia (2013)
3. ARSN (2011). http://portal.arsnorte.min-saude.pt/portal/page/portal/ARSNorte
4. Begur, S.V., Miller, D.M., Weaver, J.R.: An integrated spatial DSS for scheduling and routing home-health-care nurses. Interfaces **27**(4), 35–48 (1997)
5. Bellamy, G.R., Stone, K., Richardson, S.K., Goldsteen, R.L.: Getting from here to there: evaluating West Virginia's rural nonemergency medical transportation program. J. Rural Health **19**(S5), 397–406 (2003)
6. Berbeglia, G., Cordeau, J.F., Laporte, G.: Dynamic pickup and delivery problems. Eur. J. Oper. Res. **202**(1), 8–15 (2010)
7. Butt, S.E., Cavalier, T.M.: A heuristic for the multiple tour maximum collection problem. Comput. Oper. Res. **21**(1), 101–111 (1994)
8. Chao, I., Golden, B.L., Wasil, E.A.: A fast and effective heuristic for the orienteering problem. Eur. J. Oper. Res. **88**(3), 475–489 (1996)
9. Cordeau, J.F., Laporte, G.: The dial-a-ride problem: models and algorithms. Ann. Oper. Res. **153**(1), 29–46 (2007)
10. Czyzyk, J., Mesnier, M.P., Moré, J.J.: The NEOS server. IEEE J. Comput. Sci. Engrg. **5**, 68–75 (1998)
11. Díaz-Parra, O., Ruiz-Vanoye, J.A., Bernábe Loranca, B., Fuentes-Penna, A., Barrera-Cámara, R.A.: A survey of transportation problems. J. Appl. Math. vol. 2014, Article ID 848129, p. 17 (2014). doi:10.1155/2014/848129
12. Dolan, E.D.: NEOS server 4.0 administrative guide. Technical Memorandum ANL/MCS-TM-250, Mathematics and Computer Science Division, Argonne National Laboratory, Argonne, IL (2001)

13. DRE (2011). Despacho n.º 16843/2011, Diário da República, 2.ª série — N.º 239 — 15 de Dezembro de 2011, 48767–48768 (2011)
14. DRE (2012a). Portaria n.º 142-A/2012. Diário da República, 1.ª série — N.º 94 — 15 de maio de 2012, 2532-(2)-2532-(3) (2012)
15. DRE (2012b). Portaria n.º 142-B/2012. Diário da República, 1.ª série — N.º 94 — 15 de maio de 2012, 2532-(3)-2532-(6) (2012)
16. DRE (2012c). Declaração de Retificação n.º 27-A/2012. Diário da República, 1.ª série — N.º 107 — 1 de junho de 2012, 2928-(2) (2012)
17. DRE (2012d). Portaria nº178-B/2012. Diário da República, 1.ª série — N.º 107 — 1 de junho de 2012, 2928-(2) (2012)
18. DRE (2012e). Despacho n.º 7702-A/2012. Diário da República, 2.ª série — N.º 108 — 4 de junho de 2012, 20410-(2) (2012)
19. DRE (2012f). Despacho n.º 7702-C/2012. Diário da República, 2.ª série — N.º 108 — 4 de junho de 2012, 20410-(3)-20410-(6) (2012)
20. DRE (2012 g). Despacho nº 8705/2012. Diário da República, 2.ª série — N.º 125 — 29 de junho de 2012, 22880– 22881 (2012)
21. DRE (2012 h). Declaração de Retificação n.º 36/2012. Diário da República, 1.ª série — N.º 135 — 13 de julho de 2012, 3657– 3664 (2012)
22. Gribkovskaia, I., Laporte, G., Shyshou, A.: The single vehicle routing problem with deliveries and selective pickups. Comput. Oper. Res. **35**(9), 2908–2924 (2008)
23. Gropp, W., Moré, J.J.: Optimization environments and the NEOS server. In: Buhmann, M. D., Iserles, A., (eds.) Approximation Theory and Optimization: Tributes to M.J.D. Powell. Cambridge University Press, Cambridge, UK, pp. 167–182 (1997)
24. Gutiérrez-Jarpa, G., Marianov, V., Obreque, C.: A single vehicle routing problem with fixed delivery and optional collections. IIE Trans. **41**(12), 1067–1079 (2009)
25. Hains, I.M., Marks, A., Georgiou, A., Westbrook, J.I.: Non-emergency patient transport: what are the quality and safety issues? a systematic review. Int. J. Qual. Health Care **23**(1), 68–75 (2011)
26. Hu, Q., Lim, A.: An iterative three-component heuristic for the team orienteering problem with time windows. Eur. J. Oper. Res. **232**(2), 276–286 (2014)
27. Labadie, N., Mansini, R., Melechovský, J.: Wolfler Calvo, R.: The team orienteering problem with time windows: an lp-based granular variable neighborhood search. Eur. J. Oper. Res. **220**(1), 15–27 (2012)
28. Madsen, O.B., Ravn, H.F., Rygaard, J.M.: A heuristic algorithm for a dial-a-ride problem with time windows, multiple capacities, and multiple objectives. Ann. Oper. Res. **60**(1), 193–208 (1995)
29. Oliveira, J.A., Ferreira, J., Dias, L., Figueiredo, M., Pereira, G.: Non emergency patients transport - a mixed integer linear programming. In: Proceedings of 4th International Conference on Operations Research and Enterprise Systems (ICORES 2015), Lisboa, Portugal, ISBN 978-989-758-075-8, pp. 262–269 (2015)
30. Psaraftis, H.N.: Dynamic vehicle routing problems. In: Vehicle Routing: Methods and studies, vol. 16, pp. 223–248 (1988)
31. Safaei, J.: A ride to care–a non-emergency medical transportation service in rural British Columbia. Rural Remote Health **11**(2), 1637 (2011)
32. Stiver, H.G., Trosky, S.K., Cote, D.D., Oruck, J.L.: Self-administration of intravenous antibiotics: an efficient, cost-effective home care program. Can. Med. Assoc. J. **127**(3), 207 (1982)
33. Vaisblat, A., Albert, D.: Medical non-emergency patient centered scheduling solution. New Magenta Papers, 40 (2013)

34. Vansteenwegen, P., Souffriau, W., Oudheusden, D.V.: The orienteering problem: a survey. Eur. J. Oper. Res. **209**(1), 1–10 (2011)
35. Wilken, P., Ratnapradipa, D., Presley, D., Wodika, A.B.: An evaluation of the non-emergency medical transportation system of rural southern illinois. Am. J. Health Stud. **29**(2), 199–204 (2014)

A Simulation Study of Evaluation Heuristics for Tug Fleet Optimisation Algorithms

Robin T. Bye$^{(\boxtimes)}$ and Hans Georg Schaathun

Software and Intelligent Control Engineering Laboratory,
Faculty of Engineering and Natural Sciences,
Aalesund University College,
Postboks 1517, 6025 Ålesund, Norway
{roby,hasc}@hials.no
http://blog.hials.no/softice

Abstract. Tug fleet optimisation algorithms can be designed to solve the problem of dynamically positioning a fleet of tugs in order to mitigate the risk of oil tanker drifting accidents. In this paper, we define the 1D tug fleet optimisation problem and present a receding horizon genetic algorithm for solving it. The algorithm can be configured with a set of cost functions such that each configuration effectively constitute a unique tug fleet optimisation algorithm. To measure the performance, or merit, of such algorithms, we propose two evaluation heuristics and test them by means of a computational simulation study. Finally, we discuss our findings and some of our related work on a parallel implementation and an alternative 2D nonlinear mixed integer programming formulation of the problem.

Keywords: Receding horizon control · Genetic algorithm · Computational simulation · Dynamic optimisation · Algorithm evaluation · Modelling · Risk mitigation

1 Introduction

The marine environment along the northern Norwegian coast is constantly being exposed to the risk of drift grounding accidents of passing vessels carrying large amounts of oil or other petroleum products. To mitigate the risk of such accidents, the Norwegian Coastal Administration (NCA) runs a vessel traffic service (VTS) centre in the town of Vard. The VTS operates a fleet of tugs patrolling the coast. The task of the VTS operators is to determine suitable positional trajectories that the tugs should follow such that the risk of not saving a drifting oil tanker before it grounds is minimised.

The number of oil tanker transits is expected to rise significantly in coming years [2], therefore, the problem of commanding tugs to suitable positions

---

Robin T. Bye—This paper is an extended and revised version of a paper presented at the 4th International Conference on Operations Research and Enterprise Systems (ICORES'15) in Lisbon, Portugal, January 2015 [1].

© Springer International Publishing Switzerland 2015
D. de Werra et al. (Eds.): ICORES 2015, CCIS 577, pp. 165–190, 2015.
DOI: 10.1007/978-3-319-27680-9_11

may become unmanageable for human operators. Motivated by this challenge, researchers at the Software and Intelligent Control Engineering (SoftICE) Laboratory at the Aalesund University College (AAUC) have over the last few years developed, refined, and studied a receding horizon genetic algorithm (RHGA) [1, 3–5] through the research project Dynamic Resource Allocation with Maritime Application (DRAMA). The algorithm iteratively plans individual movement trajectories for the fleet of tugs such that the net collective behaviour of the tugs is optimised. Specifically, it employs a genetic algorithm (GA) in order to minimise cost functions that have been specifically designed to reduce the risk of drift grounding accidents.

In recent work [5], we suggested a number of cost functions that can be used with the RHGA. A challenge, however, is the problem of comparing and evaluating the merit of such cost functions, or more generally, the merit of what we have coined *tug fleet optimisation* (TFO) algorithms. This challenge is the focus of the work we present here, where we propose two new and objective evaluation heuristics designed for making comparisons of TFO algorithms.

In the following sections, we give a brief but up-to-date review of relevant traffic statistics and motivation for our work before presenting a 1D model of the TFO problem. The RHGA is described briefly and a set of possible cost functions that can be used to configure the algorithm is provided. We then propose two new and objective evaluation heuristics designed for measuring and comparing the merit of TFO algorithms. Details about our simulator framework and its implementation in the functional programming language Haskell are given in a separate section. The RHGA and the set of cost functions are tested by means of a computational simulation study on a large number of simulation scenarios, in which the performance of each RHGA configuration is measured by means of the proposed evaluation heuristics. Finally, we discuss our findings and some of our related work on parallelisation and an alternative 2D nonlinear mixed integer programming formulation of the problem.

2   Method

2.1   Background and Motivation

Several thousand ships transit along the northern Norwegian coastline every year. The Vard VTS, which is located at the northeasternmost point of Norway (see Fig. 1), constantly monitors ship movements, maintains dialogue with ships, and manages the tug fleet of Norway. With the recent increase in traffic through the Northwest Passage and the projected increase in oil exploration in the High North [2], the Norwegian coastline is increasingly exposed to the risk of incidents with potentially high impact on the environment. The latest figures for 2014 have not been released but for 2013 consisted of 1,584 so-called "risky transports," of which 298 were ships with oil or other petroleum-related cargo on board [7]. In 2014 alone, the Vard VTS registered 298 operational incidents, including 167 incidents of drifting vessels, 31 of grounding, 51 of pollution, 8 of fire, and 13 shipwrecks [8]. In 2013, the same figures were 286 operational incidents, including

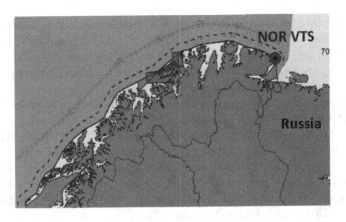

Fig. 1. Northern Norwegian coastline and the Vard VTS (shown with its call signal NOR VTS). Solid line is the geographical baseline; stapled line is the border of the Norwegian Territorial Waters (NTW); thick pink line is the traffic corridor for the Traffic Separation Scheme (TSS). Adapted from [6] (Color figure online).

186 incidents of drifting vessels, 29 of grounding, 36 of pollution, 10 of fire, and 7 shipwrecks [9].

As noted above, there is an incident of a drifting vessel occurring about every second day on average. A number of these vessels are high-risk ships such as oil tankers, which if allowed to drift aground can cause serious damage to the environment due to spillage of oil and fuel. In a measure to avoid such incidents, the VTS is constantly instructing its patrolling fleet of tugs to move to new positions in a manner such that if an oil tanker loses manoeuvrability, e.g., because of engine or propulsion problems or steering failure, tugs should be sufficiently close that it can intercept the drifting oil tanker before it runs aground [10].

A set of risk-based decision support tools based on dynamical risk models have been developed previously [10,11]. The models incorporate a number of factors such as wind, waves, currents, geography, types of ships in transit, and potential environmental impact should drift grounding occur. Whilst such tools can aid the human operators at the VTS in making informed decisions about tug fleet positioning, they do not tell explicitly where tugs should move; instead, they give the operators a real-time risk map divided into zones of low, medium, and high risk. With the expected increase in oil tanker transits [2], the problem of commanding tugs to suitable positions may become unmanageable for human operators and motivates the need for TFO algorithms.

2.2   A 1D Model of the Tug Fleet Optimisation Problem

We employ a 1D model of the TFO problem that adopts most of the principles and assumptions in our earlier work [3–5,12]. For more details and justification of assumptions, we refer to [1].

Oil tankers are required by law to follow a predefined corridor, or lane, parallel to the coastline, depicted as the pink TSS in Fig. 1. For simplicity, we assume that N_o oil tankers move in one dimension only along a straight line of motion z, whereas N_p tugs are patrolling along a 1,500 km long line of motion y parallel to z, e.g., the geographical baseline depicted in Fig. 1.

For any oil tanker moving along the line z, there is a small probability that an incident may occur at the position $z(t)$, resulting in the tanker starting to drift at $t = t_d$. Naturally, most of the time, nothing will happen, and the tanker will continue sailing along z. Employing a discrete-time model with a sampling period of $t_s = 1$ h, we assume that we can estimate the future tanker positions at discrete points in time, limited to a *prediction horizon* T_h hours into the future. For each of the oil tankers, this results in a set of future tanker positions given by $\{\hat{z}(t|t_d)\}$ for $t = t_d + 1, t_d + 2, \ldots, t_d + T_h$.

Furthermore, we assume that we can determine, for example through Monte Carlo simulations, the most likely hypothetical predicted drift trajectories that emanate from each predicted tanker position $\hat{z}(t|t_d)$. Such trajectories depend on a number of actual and forecast conditions in the area, such as ocean currents and wind speed and direction, and may or may not intersect the patrol line y after an *estimated drift duration* $\hat{\Delta}$ into the future. Realistic drift trajectory models exist and are currently an active focus of research (e.g., see [10]).

According to [10], situations of "fast drift" can have drift durations as fast as 8–12 h, whereas more typical drift durations are in the range 16–24 h. In previous work, in order to be conservative rather than optimistic, we therefore either set the estimated drift duration $\hat{\Delta}$ to be 8 h for all oil tankers [12], or to be drawn randomly for each oil tanker such that $\hat{\Delta} \in \{8, 9 \ldots, 12\}$ hours [3–5].

It should also be kept in mind that there will inevitably be a *detection delay* δ between the time when an oil tanker begins drifting at the *drift time* t_d[1] and the time when the VTS centre detects, or is notified of, the incident at time t_a some hours later, which we call the *alarm time*. The detection delay is thus given by $\delta = t_a - t_d$.

If we examine all the future predicted positions for all the oil tankers as well as all the corresponding drift trajectories, we obtain a distribution of *cross points* located at points where future drift trajectories will intersect the patrol line y. A cross point of the cth oil tanker's drift trajectory at time t can be defined as the position y_t^c. Assuming a drift duration $\hat{\Delta}$, a drift trajectory starting on $z(t)$ at $t = t_d$ will have a cross point on y at $t = t_d + \hat{\Delta}$. Assuming the same drift duration for all drift trajectories and considering the prediction horizon T_h, there is a predicted set of cross points for the cth oil tanker given by

$$\{y_t^c\} = \left\{ y_{t_d+\hat{\Delta}}^c, y_{t_d+1+\hat{\Delta}}^c, \ldots, y_{t_d+T_h}^c \right\}. \tag{1}$$

Moreover, we define a *patrol point* as the pth tug's position on y at time t as y_t^p.

Based on the predicted future distribution of cross points, we define the TFO problem as the problem of calculating patrol trajectories (sequences of patrol

---

[1] Note that t_d also is used as the start time for planning patrol trajectories for the tugs to follow.

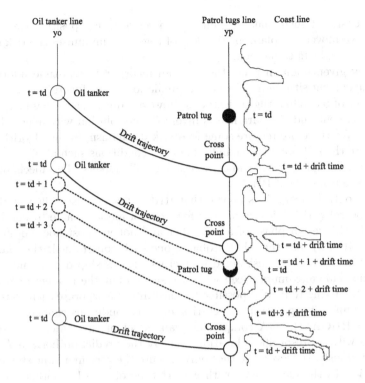

Fig. 2. Tug fleet optimisation problem: where should the tugs move?

points) that start at $t = t_d$ and have some duration T_h, along y for each of the patrolling tugs such that the risk of an oil tanker in drift not being reached and prevented from grounding is minimised.

Figure 2 shows a graphical summary of the TFO problem as presented above, exemplified by two patrolling tugs and three oil tankers.

2.3   The Receding Horizon Genetic Algorithm

The TFO algorithm that we study in this paper is the RHGA [3–5]. The algorithm consists of two main components: receding horizon control (RHC) and a genetic algorithm (GA). The GA is a search heuristic for solving search and optimisation problems and is inspired by elements in natural evolution, such as inheritance, mutation, selection, and crossover. It has been attributed to [13], with subsequent popularisation by [14], and is currently a very popular optimisation tool across many different disciplines, including operations research. The GA we have implemented in our RHGA is roughly follows the outline of GAs presented in [15].

The optimisation problem must be defined as a cost function such that, when evaluated for a set of candidate solutions, the GA is able to distinguish good solutions from bad ones. Specifically, for the TFO problem, the cost function must be designed such that its solution is a set of future position trajectories, or collective movement plan, for the fleet of tugs that minimises the risk of drift grounding accidents to happen.

At any given point in time, the GA can incorporate real-time information about the current situation, as well as a prediction of the future, to calculate an optimal set of patrolling tug trajectories. However, due to the dynamic nature of the environment and the parameters involved, the solution will quickly become outdated. We therefore require some feedback mechanism in the algorithm that can update the solution with changes in ocean conditions such as wind, current, and waves, as well as speed and direction of oil tankers. The mechanism we adopt is the principle of RHC.

From control theory, it is known that RHC, which is also called model predictive control (MPC), is one of very few control methods able to handle constraints in the design phase of a controller and not via post hoc modifications (e.g., see [16–18]). For the TFO problem, one such constraint is the maximum speed of tugs, which is constrained by factors such as ship design and weather conditions. This maximum speed will necessarily limit the number of *reachable cross points*. Using RHC it is possible to constantly incorporate such constraints in the planning of tug patrol trajectories, even as conditions change.

In our RHGA, the GA component plans a set of tug trajectories starting at t_d and with a prespecified duration, namely the prediction horizon T_h introduced previously. However, the tugs only execute the very first time step of their trajectories. In the mean time, with a start time of $t_d + 1$, another set of tug trajectories is planned, based on new and predicted information available. This new solution replaces the old one but again only the first portion is implemented. This process repeats as a sequence of planning steps, thus creating a feedback loop where updated information is fed back to the GA. Effectively, the prediction horizon keeps being shifted into the future, and this has led to the term receding horizon control.

2.4   Cost Functions

Determining suitable cost functions for a TFO algorithm is a key design challenge for the algorithm to be successful. Below we present three possible main cost functions, each of which can be configured to yield different properties by means of parameterisation. For more detailed explanations of the cost functions, we refer to [1].

Cost Function f_1. We employ a metric from our earliest work [3,4,12] defined as the sum of the distances between all cross points and the *nearest* patrol points, based on the argument that if an oil tanker in drift can be saved by the nearest tug, then it is not relevant if the other tugs are also able to save the tanker. As introduced in [5], we also include the options of a "power" parameter e and raise the distance to the power $e = 2$ for squaring distances (cross points further away will be penalised more), whereas we use a safe region parameter r to impose zero penalty in the cost function for close cross points. Note that letting $e = 1$ and $r = 0$ yields the cost function used in our earliest work [3,4,12].

Cost function f_1 is then given by

$$f_1(t) = \sum_{t=t_d}^{t_d+T_h} \sum_{o \in O} \max \left\{ 0, \min_{p \in P} |y_t^c - y_t^p|^e - r \right\} \tag{2}$$

for N_o oil tankers $o \in O = \{o_1, \ldots, o_{N_o}\}$, N_p patrol tugs $p \in P = \{p_1, \ldots, p_{N_p}\}$, $e \in \{1, 2\}$, and r chosen as some distance that can confidently be reached by a tug to enable drift interception and hookup to the ship. A reasonable and conservative choice for r could for instance be half the expected distance a tug can travel from an alarm is received until the first hypothetical cross points occur.

Cost Function f_2. In [5] we show that a detection delay $\delta = t_a - t_d$ must be taken into account in the cost function, where t_d is the time when an oil tanker starts drifting and t_a is the time when the VTS is alarmed some hours later. We also define a new, and shorter, drift-from-alarm (DFA) time $\hat{\Delta}_a = \hat{\Delta} - \delta$. We use a somewhat arbitrary, but realistic, choice of $\delta = 3\,\text{h}$ [1,5]. Moreover, the cost function should compare the positions of the tugs (patrol points) *when* they receive the alarm at time t_a, and the hypothetical future positions where drifting tankers will cross the patrol line some $\hat{\Delta}_a$ hours later, where $\hat{\Delta}_a$ is the total drift time $\hat{\Delta}$ (8–12 h) less the detection delay δ (3 h), leaving $\hat{\Delta}_a$ in the range 5–9 h.

The modified cost function f_2 is then given by

$$f_2(t) = \sum_{t=t_a}^{t_a+T_h} \sum_{o \in O} \max \left\{ 0, \min_{p \in P} \left| y_{t+\hat{\Delta}_a}^c - y_t^p \right|^e - r \right\}. \tag{3}$$

For further details, we refer to [1].

Cost Function f_3. A third possible measure is the number of unsalvageable tankers. That is, we merely consider whether a tug can reach a drifting tanker in time to prevent it from grounding, and the distance is otherwise immaterial. We may use the safe range r for counting the number of unreachable cross points and let this number constitute a measure for unsalvageable tankers. If cross points are outside the safe range, we add 1 to the accumulated cost, otherwise we add 0. The cost function f_3 can then be described by

$$f_3(t) = \sum_{t=t_a}^{t_a+T_h} \sum_{o \in O} g \left(\min_{p \in P} \left| y_{t+\hat{\Delta}_a}^c - y_t^p \right| - r \right), \tag{4}$$

$$g(x) = \begin{cases} 1, & x > 0 \quad (\text{outside } r), \\ 0, & x \le 0 \quad (\text{inside } r), \end{cases} \tag{5}$$

where $g(x)$ is the Heaviside unit step function.

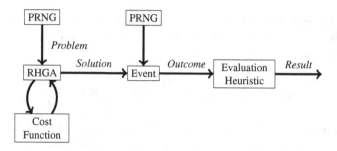

Fig. 3. Simulation model. A pseudo-random number generator (PRNG) generates simulation scenarios in which the RHGA uses some cost function Cost 1 to determine a solution (where tugs should move). The PRNG then generates an event of drift, which depending on the current position of the tugs may be critical or not. The outcome (saving or not saving the drifting ship) is then quantified as Cost 2 by an evaluation heuristic.

2.5   Evaluating Algorithm Performance

How can we compare the performance of different TFO algorithms, or in our case, the performance of the RHGA configured with various different cost functions? By definition, the metrics of different cost functions are not generally directly comparable, and it is not always possible to use a cost function which directly reflects the real cost of the solution. In the TFO problem, there are many random elements without well-understood probability models. Incorporating these elements in the cost function would make it too complex to be practical.

Simulation Model. Our proposed solution is a Monte Carlo simulation as shown in Fig. 3, where the complete optimisation algorithm with cost function can be tested against a large number of (pseudo) random scenarios.

We have two pseudo-random algorithms (PRNG); one to generate the problem as observed by the optimisation algorithm (RHGA), and one to generate the situation, or *event*, where the solution is to be executed. For example, the positions of the tankers is known *a priori*, and is part of the problem. A tanker starting to drift is an event which is only known after the RHGA has provided the solution. Given the solution and the event, we can evaluate the cost of the result, e.g., of a grounding accident or a successful rescue.

The evaluation heuristic may look similar to the cost function, but there is a critical difference. The evaluation heuristic evaluates the cost of a particular event. The cost function, on the other hand, has to evaluate a solution without the knowledge of which event will occur.

There are several stochastic processes governing the outcome of an event in the model. These processes can be internalised either in the Monte Carlo simulation or integrated analytically in the evaluation heuristic. We can illustrate this with an example. The event in the simulation model can be subdivided into two stages:

1. Oil tanker o starts drifting.
2. Oil tanker o grounds.

If the first event occurs, one ore more patrol tugs will attempt to rescue the drifting tanker. This rescue operation may or may not succeed, depending largely on the maximum tug speed as determined by weather conditions. If it does not, the second event occurs.

One possibility is to stop the Monte Carlo estimation after the first event, and let the cost be equal to the conditional probability of the second event, that is, the probability of a failed rescue operation. The second possibility is to simulate the entire rescue operation. If it succeeds the cost is zero, otherwise it is the cost of the grounding accident, which may depend on the type of cargo, geographical location, weather conditions, and so on.

For the purpose of this work, we evaluate the result when tanker o starts drifting. We do not simulate the rescue operation. The steps of the evaluation method can be summarised as follows:

1. randomly generate a deterministic and reproducible simulation scenario;
2. run the RHGA (or another TFO algorithm) for a given number of planning steps;
3. considering each oil tanker separately, assume each tanker begins drifting and count the number of salvageable tankers;
4. for the same simulation scenario, repeat (2) and (3) with a different cost function configuration in the RHGA (or a different TFO algorithm); and
5. repeat steps (1)–(4) for a number of different simulation scenarios and find the accumulated evaluation cost for each RHGA configuration (or TFO algorithm).

Note that instead of evaluating one random event, we evaluate one event for each tanker o, where o starts to drift. This is possible because the number of tankers is small, and it lets us evaluate a larger number of scenarios with little extra time. We propose two candidate evaluation heuristics h_1 and h_2.

Evaluation Heuristic h_1. The first heuristic is similar to cost function f_3 counting the number of salvageable tankers at some alarm time t_a. We will simply assume that each patrol tug p can save any ship with cross points inside the safe region $r = v_{\max}^p \hat{\Delta}_a$ away, where $\hat{\Delta}_a$ is the DFA time and $v_{\max}^p$ is the pth tug's maximum speed, that is, within the maximal reach of a tug upon a drift alarm. In a more realistic model, the heuristic should probably be weather dependent and direction dependent, e.g., going against the wind is slower than going with it. It should also include hookup times.

A simulation scenario in this case is simply a set of pre-determined oil tanker movements and the resulting hypothetical drift trajectories and cross points for a pre-specified duration. For testing purposes, we can generate a number of such scenarios offline and use them as input data for testing TFO algorithms. In a real-world application, the actual scenario is what that is happening right now, and future oil tanker positions, drift trajectories, and cross points would have to be predicted in real-time.

To sum up, we define the evaluation heuristic h_1 as

$$h_1(t_a) = \sum_{o \in O} g\left(\min_{p \in P}\left|y^c_{t_a+\hat{\Delta}_a} - y^p_{t_a}\right| - r\right),\tag{6}$$

$$r = v^p_{\max}\hat{\Delta}_a\tag{7}$$

$$g(x) = \begin{cases} 1, & x > 0 \quad \text{(outside } r\text{),} \\ 0, & x \leq 0 \quad \text{(inside } r\text{).} \end{cases}\tag{8}$$

Other possible objective measures exist, e.g., we could sum up the total fuel consumption and use it as a component of an overall objective measure if that is of interest. Furthermore, the cost of measuring the number of salvageable tankers does not need to be discrete (yes/no) but could instead have a continuous probability distribution attached to it. We could then sum these probabilities to find an evaluation cost for the TFO algorithm.

Evaluation Heuristic h_2. The evaluation heuristic h_1 does not discriminate between cross points far away from the nearest tug and cross points that are much closer, as long as they are all inside the maximal reach from any tug as given by $r = v^p_{\max}\hat{\Delta}_a$. However, it is clear that due to varying and non-optimal weather conditions, the maximum speed of each tug may be much lower than during ordinary operation. Moreover, h_1 does not take into account that there will be a hookup time when the tug attaches itself to the drifting ship. In an attempt to address these issues, we suggest the following evaluation heuristic h_2 given by

$$h_2(t_a) = \sum_{o \in O}\left(\max\left\{0, \min_{p \in P}\left|y^c_{t_a+\hat{\Delta}_a} - y^p_{t_a}\right| - r\right\}\right)^2,\tag{9}$$

$$r = v^p_{\min}\hat{\Delta}_a,\tag{10}$$

where the safe region has been reduced to the area reachable for any tug with some minimum speed $v^p_{\min}$, which we assume the tug will always be able to maintain. Inside the safe region, there is zero cost for cross points of salvageable tankers, whereas outside, the cost increases with the square of the distance to cross points of unsalvageable tankers. Squaring ensures that we punish larger distances more.

2.6   Simulator Framework

Whilst the RHGA was implemented in MATLAB in earlier versions [3,4], we recently rewrote the entire code base in the advanced, purely-functional programming language Haskell[2] for our latest work [1,5].

---

[2] http://www.haskell.org.

A thorough presentation of our new simulator framework requires a separate paper but we will cover the essential below. We chose the purely-functional programming language Haskell for our implementation, which means that functions in Haskell are pure, there is no global state, and no side effects. Code written in Haskell is therefore less error-prone and usually more concise, compact, and readable than imperative programming languages like C or Java. An additional advantage that "comes for free" with a functional language is a focus on *what* the programmer wants to achieve, rather than *how*, since functional program specifications can simply be executed directly rather than translated into imperative code. Moreover, whilst being strongly typed, and thus avoiding compile-time core dumps, Haskell uses polymorphism, which enhances the reusability of code. Many functions can therefore be written only once, because they accept input and output variables of many different types.

Haskell is also a good choice for parallel programming, which we believe is likely to be needed as the complexity of our simulator grows. Using pure parallelism guarantees deterministic processes and zero race conditions or deadlocks, however, as we describe in more detail in the discussion (see Sect. 5.2), parallelisation of pseudo-random number generators (PRNGs) represents a challenge since, by definition, PRNGs are impure and require book-keeping of a system state. Nevertheless, using a functional language like Haskell provides a clear separation between pure and impure functionality, thus reducing this book-keeping to a minimum.

Finally, it is worth mentioning that Haskell is a non-strict, lazy language, meaning that evaluation only happens on demand. This removes the need for the programmer to pre-allocate memory such as fixed-size arrays, and makes it easier to write modular programs, since functions can be passed freely to other functions, be returned as the result of a function, and stored in data structures.

Our choice of using Haskell for implementation makes our simulator modular and easily extendable and we are therefore confident that we will be able to perform several comprehensive and quantitative studies in the time to come.

3 Computational Simulation Study

3.1 Basic Parameters

In previous papers, we have consistently assumed $N_p = 3$ tugs in the fleet. Using our evaluation heuristics, we are able to examine by means of simulations the difference in coastal protection dependent on fleet size. In order to test our propositioned evaluation heuristics, we therefore let $N_p \in \{1, \ldots, 6\}$. Moreover, based on a desire to obtain results comparable with our previous work, we set the number of tankers sailing along z and being watched by the Vard VTS to a constant $N_o = 6$.

The stretch of coastline patrolled by the tug vessels that we have termed the patrol line y is about $1,500$ km long. Hence, we define our patrol zone Y as unidimensional along y in the continuous interval $Y = [-750, 750]$ km, and constrain cross points y^c and patrol points y^p to lie in Y, or $y^c, y^p \in Y$. For implementation purposes, we also define a tanker zone Z in the same interval much further away from land but underline that the (simulated) VTS will still observe ship traffic outside of this zone. The reason is of course that a tanker outside the tanker zone Z may still drift and ground inside the patrol zone Y.

Under normal conditions, we assume that the patrol tugs are limited to a maximum speed of $v_{\max}^p = 20$ km/h, whereas the speed of each oil tanker v^o is randomly drawn from a uniform distribution such that $v^o \in [20, 30]$ km/h. Note that compared with previous work, we have reduced the maximum speed of the tugs from 30 km/h to 20 km/h, thus making it more difficult for tugs to cover potential cross points. These speeds are in line with the literature (e.g., [10, 19]). We also assume that even in very bad weather conditions, the tugs are able to maintain at least a minimum speed of $v_{\min}^p = 5$ km/h.

Drift trajectories are set to be perpendicular (eastbound) onto the south-north patrol line y, with associated drift times drawn randomly from the interval $\{8, 9, \ldots, 12\}$ hours and cross points generated from extrapolating the predicted future positions of oil tankers and their resulting drift trajectories.

The parameter settings are summarised in Table 1.

3.2   Simulation Scenarios

A simulation scenario consists of simulation-generated tanker movements along z as well as hypothetical drift trajectories with corresponding cross points on y. The scenario acts as an input to a TFO algorithm such as the RHGA (see Fig. 3) and is completely independent of what the RHGA calculates and how the tugs move.

We initialise a scenario by placing N_o oil tankers at random positions and with random speeds along z, headed in either the southbound or the northbound direction. Next, we sample each of the tankers' positions, speeds, and directions at every simulation step $t_s = 1$ h from the start of the simulation at $t_i = 0$ h to the final simulation time at $t_f = 24$ h. For any simulation time t_d in $\{t_i, t_i + t_s, \ldots, t_f\}$, we suppose that we have precise real-time information about the speed and direction of each oil tanker, as provided by AIS. We also assume that we have an accurate model that, given this real-time actual information, is able to predict future positions and speeds of the tankers at future times $t_d + t_s, t_d + 2t_s, \ldots, t_d + T_h$, where T_h is the prediction horizon. Finally, we assume that we have another accurate model that is able to predict hypothetical drift trajectories and cross points for each tanker if it starts drifting at time t_d and also at the future times just listed.

Note that since the scenario consisting of oil tanker movements, drift trajectories, and cross points is independent of how the fleet of tugs move, we can replay the same scenario as an input to other TFO algorithms (or variations of the RHGA) in order to evaluate and compare the algorithms.

Table 1. Simulation parameters, settings, and units.

Parameters	Settings	Units
Patrol zone (south-north line)	$Y = [-750, 750]$	km
Tanker zone (south-north line)	$Z = [-750, 750]$	km
Number of oil tankers	$N_o = 6$	-
Set of oil tankers	$O = \{1, 2 \dots, N_o\}$	-
Number of tugs	$N_p = \{1, \dots, 6\}$	-
Set of tugs	$P = \{1, 2 \dots, N_p\}$	-
Initial tug positions (base stations)	Uniformly distributed	km
Random initial tanker positions	$y^o \in Z, \forall o \in O$	km
Maximum speed of tugs	$v_{max}^p = 20, \ \forall p \in P$	km/h
Minimum speed of tugs	$v_{min}^p = 5, \ \forall p \in P$	km/h
Random speed of oil tankers	$v^o \in [20, 30], \forall o \in O$	km/h
Initial simulation time	$t_i = 0$	h
Simulation step	$t_s = 1$	h
Final simulation time	$t_f = 24$	h
Prediction horizon	$T_h = 24$	h
Time of start of drift	$t_d \in \{t_i, t_i + 1, \dots, t_f\}$	h
Detection delay	$\delta = 3$	h
Alarm time	$t_a = t_d + \delta \in \{t_i + \delta, t_i + \delta + 1, \dots, t_f + \delta\}$	h
Drift direction	Eastbound	-
Estimated drift times	$\hat{\Delta} \in \{8, 9, \dots, 12\}$	h
Drift-from-alarm (DFA) times	$\hat{\Delta}_a = \hat{\Delta} - \delta \in \{5, 6, \dots, 9\}$	h
Static strategy	$y_t^p = y_{t_i}^p, \ \forall t$	km
Cost functions	$F = \{f_1, f_2, f_3\}$	-
Distance power	$e = \{1, 2\}, \quad \text{in} \quad f_1, f_2$	-
Safe region	$r = \begin{cases} \{0, 50, 100\}, & \text{in} \quad f_1, f_2 \\ \{50, 100\}, & \text{in} \quad f_3 \\ v_{max}^p \hat{\Delta}_a = [100, 180], & \text{in} \quad h_1 \\ v_{min}^p \hat{\Delta}_a = [25, 45], & \text{in} \quad h_2 \end{cases}$	km
TFO algorithms	Configurations of $\text{RHGA}(f_i, e, r, N_p)$	-
Number of $\text{RHGA}(f_i, e, r, N_p)$ configurations	$N_{conf} = 15$	-
Number of scenarios	$N_{sc} = 1600$	-
Total number of simulations	$N_{sim} = N_{conf} \times N_{sc} \times \dim N_p = 144,000$	-

3.3 Tug Fleet Optimisation Using the Receding Horizon Genetic Algorithm

For all scenarios, the N_p tugs are initialised at simulation time $t_i = 0$ by being uniformly positioned along the coast at stationary base stations in a manner such that they can cover as much of the patrol line y as possible. For example, since we have defined y as a line constrained to $[-750, 750]$ km, a single tug will be placed at $y_{t_i}^p = 0$, a fleet of two tugs will be placed at $y_{t_i}^p = \{-375, 375\}$, a fleet of three tugs will be placed at $y_{t_i}^p = \{-500, 0, 500\}$, and so on. From these initial positions, tugs will begin to actively pursue good positions for reducing the risk of drift grounding accidents depending on how the scenario plays out and how the TFO algorithm will control them.

At any simulation time t_d, the GA component in the RHGA uses the predicted distribution of potential cross points to calculate a plan, which consists of a position trajectory for each of the tugs in the fleet. The plan consists of future desired positions for each tug at times $\{t_d + 1, t_d + 2, \ldots, t_d + T_h\}$. The plan is optimal (or close to optimal) in the sense that it minimises (or tries to minimise) a cost function.

Using RHC, we let the tugs execute only the first step of this plan and move the tugs from their positions at $t = t_d$ to future positions at $t = t_d + 1$. At $t = t_d + 1$, the GA plans a new set of desired trajectories from $t = t_d + 2$ to $t = t_d + T_h + 1$, but again, we let the tugs execute only the first step from $t = t_d + 1$ to $t = t_d + 2$. This process repeats until the final simulation time $t_f = 24$, at which we plan for $t = t_f, t_f + 1, \ldots, t_f + T_h$, and again, and finally, let the tugs execute only the first step from $t = t_f$ to $t = t_f + 1$.

We have then completed one simulation of this particular scenario using one particular TFO algorithm, in our case, the RHGA employing a particular configuration of one of the cost functions f_1–f_3. The end-of-simulation positions of tugs and tankers and their cross points are then be used as input to the evaluation heuristics.

3.4   Settings of the Genetic Algorithm

The GA we employ in this study is loosely based on the GAs presented in [15] and has been presented in detail in our earlier work [3,4]. We initialise the GA with a population size of chromosomes that are randomly generated. At every iteration of the GA, $N_{keep} = 10$ chromosomes are selected from the population by roulette wheel selection, with low cost chromosomes having a greater chance of being picked. These chromosomes survive from one generation to the next and are also used for mating to generate new offspring that replace the chromosomes that were not picked. Mating is performed by a combination of an extrapolation method and a single crossover point to obtain new offspring variable values bracketed by the parents variable values (see [15]) for details. After mating, the new population of chromosomes is ranked and the $N_{elite} = 10$ best chromosomes are categorised at elite chromosomes and are not allowed to mutate. Of the remaining non-elite chromosomes, each has a mutation rate $\mu = 0.1$ probability of being mutated. After mutation has taken place, the GA repeats the process for a total of $N_{iter} = 200$ iterations, after which the best solution obtained is used for moving the tugs one step ahead as per the RHC strategy presented above.

We chose these GA parameter settings by manually evaluating a number of test runs, where we were able to find suitable settings that ensured satisfactory run times for all RHGA configurations while at the same time obtaining satisfactory minimisation of cost functions. That is, only negligible improvements were attainable from tuning the GA to other, often more time-consuming settings, e.g., increasing the population size or number of iterations.

Table 2. GA settings.

Parameters	Settings
Population size	$N_{\mathrm{pop}} = 50$
Chromosomes kept for mating	$N_{\mathrm{keep}} = 10$
Elite chromosomes	$N_{\mathrm{elite}} = 10$
Mutation rate	$\mu = 0.1$
Number of iterations	$N_{\mathrm{iter}} = 200$

Our choice of GA settings are summarised in Table 2.

3.5 Configurations of the Receding Horizon Genetic Algorithm

The three cost functions f_1, f_2, and f_3 are parameterised by a distance power e and a safe region r, whereas the evaluation heuristics h_1 and h_2 are parameterised by a safe region r, which in turn is a function of v_{max}^p or v_{min}^p for h_1 or h_2, respectively.

We decided to implement and evaluate 14 different configurations of the cost functions for the RHGA and evaluate each configuration using both the evaluation heuristics. In addition, we wanted to evaluate a static strategy, in which tugs are stationary at base stations and do not move until notified about a drifting ship. For convenience in our data processing, we have labelled the static strategy as configuration #0 and defined its configuration as f_0, with e and r both set to zero. For the same reason, we have set $e = 0$ for f_3, where e is not applicable. Each configuration can be thought of as a unique TFO algorithm. Indeed, our approach generalises to the evaluation of any TFO algorithm able to calculate tug fleet control decisions based on the parameters, settings, and input scenarios that we have described above.

The 15 configurations are summarised in Table 3.

4 Results

We randomly generated $N_{\mathrm{sim}} = 1600$ unique simulation scenarios and tested the performance of tug fleet optimisation for each of the 15 RHGA configurations given in Table 3 when faced with $N_{\mathrm{p}} = \{1, \ldots, 6\}$ tugs to control, yielding a grand total of 144,000 simulations. For each configuration, that is, each combination of cost function f_i, distance power e, and safe region r, denoted as $f_i(e, r)$, the sample mean, standard deviation, coefficient of variance (relative standard deviation), standard error (standard deviation of the sample mean), and relative standard error for the evaluation heuristics h_1 and h_2 were calculated for $N_{\mathrm{p}} = \{1, \ldots, 6\}$ tugs, respectively. Both heuristics were evaluated at the end of each simulated scenario at $t = t_{\mathrm{f}}$. In the sections below, the results of the sample means as well as comparisons of the active control configurations of the RHGA versus the static strategy are our main concern and will be presented graphically, whereas the other statistics will be presented briefly in text.

Table 3. RHGA configurations.

Cost function f_i	Power e	Safe region r	#
0	0	0	1
1	1	0	2
		50	3
		100	4
	2	0	5
		50	6
		100	7
2	1	0	8
		50	9
		100	10
	2	0	11
		50	12
		100	13
3	0	50	14
		100	15

4.1   Evaluation Heuristic h_1

The evaluation heuristic h_1 is a measure of the number of unsalvageable tankers. Figure 4 (top) shows the maximum (worst performance) and minimum (best performance) sample mean $\bar{h}_1$ of cost functions f_1–f_3 over all configurations (combinations of power e and safe region r), as well as the static strategy, evaluated for 1–6 tugs.

Unsurprisingly, the size of the tug fleet strongly affects h_1. With a single tug, $\bar{h}_1$ was in the range [4.15–4.87], and then decreased with the number of tugs to the range [0.034–0.30] for six tugs.

For all configurations with 1–3 tugs, the standard deviation showed no trend and was in the range [0.99–1.25], whereas it was decreasing with the number of tugs for configurations with 4–6 tugs and ranged from 0.19 to 0.91. The standard error of the mean was in the range [0.005–0.032] for all configurations, with typically smaller values for smaller means. The relative standard error (found by dividing by the mean) was small for all configurations, increased with number of tugs (and thus smaller means), and was in the range [0.0051–0.13].

As expected, all RHGA configurations of f_1–f_3 outperform the static strategy for the same number of tugs. In addition, when the RHGA is configured with a tug fleet with one tug less than the static strategy, the following observations are made: With a single tug, no RHGA configuration is able to outperform the static strategy with two tugs; with two tugs, the best configurations of f_2 and f_3 outperform the static strategy with three tugs; and with 3–5 tugs, all

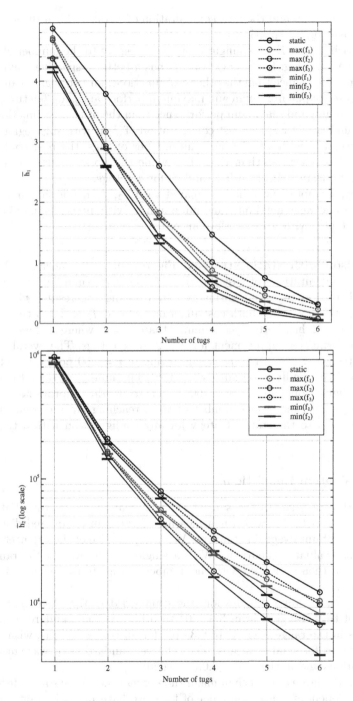

Fig. 4. Maximum and minimum mean of h_1 (top) and h_2 (bottom) evaluated for 1–6 tugs and cost functions f_1–f_3 and the static strategy (Color figure online).

configurations of f_2 and the best configuration of f_3 outperform the static strategy with 4–6 tugs, respectively.

Comparing the RHGA configurations with respect to the number of tugs, we observe the following: For a single tug, f_3 has a better respective minimum and maximum performance than the other cost functions; for two tugs, f_2 and f_3 have approximately equal minimum and maximum performance; for 3–6 tugs, f_2 has a better minimum and maximum performance than the other cost functions. Also, for any number of tugs, the best configuration of f_3 is with safe region $r = 100$ and the worst is with $r = 50$. For any number of tugs, the best configuration of f_3 is not much worse than that of f_2 while the worst configuration of f_3 is clearly worse than that of both f_1 and f_2 for 4–6 tugs.

Finally, we note that when compared with f_2, f_1 has a similar trend and relationship between maximum and minimum $\bar{h}_1$ with increasing number of tugs but consistently with worse performance.

Comparison with Static Strategy. Figure 5 shows the normalised mean of h_1 for all configurations of cost functions f_1 (left), f_2 (middle), and f_3 (right) evaluated for 1–6 tugs and normalised by dividing the results with those of the static strategy. For both cost functions f_1 and f_2 and 1–3 tugs, a power setting of $e = 1$ has better performance than $e = 2$, whereas there is a slight overall performance improvement for $e = 2$ for 4–6 tugs. The overall best safe region setting is $r = 50$. For f_3, a safe region of $r = 100$ performs well, whilst $r = 50$ performs badly, especially with increasing number of tugs. Finally, only f_2 (all configurations) and f_3 ($r = 100$) is able to steadily improve its normalised performance with increasing number of tugs, reaching an improvement of 73–88 % compared to the static strategy for any configuration with a tug fleet of six tugs.

4.2   Evaluation Heuristic h_2

Figure 4 (bottom) shows the same results as that of Fig. 4 (top) but for the evaluation heuristic h_2, which is a measure of the sum of squared distances to cross points of unsalvageable tankers. As for h_1, the size of the tug fleet strongly affects this evaluation heuristic. With a single tug, $\bar{h}_2$ was in the range $[8.5–9.7] \cdot 10^5$, and then decreased with the number of tugs to the range $[3.7–9.5] \cdot 10^3$ for six tugs.

Generally, for any configuration, the standard deviation decreased with the number of tugs, ranging from $3.3 \cdot 10^3$ to $5.1 \cdot 10^5$. The standard error of the mean was in the range from 82 to $9.3 \cdot 10^3$ for all configurations, with typically smaller values for smaller means. The relative standard error was small for all configurations and in the range $[0.0091–0.026]$.

Due to the large difference in magnitude of $\bar{h}_2$ (due to the square term in the heuristic) depending on the number of tugs, we have plotted $\bar{h}_2$ on a logarithmic scale to enhance readability. The results are similar to those for evaluation heuristic h_1, with the same relationships between the various cost functions and

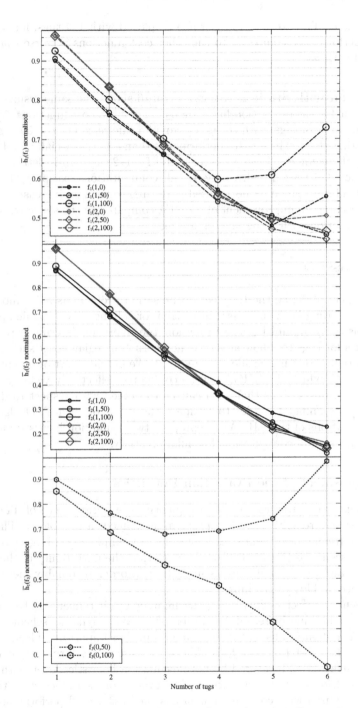

Fig. 5. Mean h_1 for cost functions f_1 (top), f_2 (middle), and f_3 (bottom), normalised by the static strategy (Color figure online).

configurations. The exception is f_3 when employed with 1–4 tugs, for which its relative performance compared to the other cost functions is worse than when evaluated with h_1.

Comparison with Static Strategy. Figure 6 shows the same results as that of Fig. 5 but for the evaluation heuristic h_2. For both cost functions f_1 and f_2 and any number of tugs, with the exception of a safe region of $r = 50$, a squared power setting of $e = 2$ has better performance than $e = 1$, especially for 4–6 tugs. For f_3, a safe region of $r = 50$ performs better for 1–2 tugs, whilst letting $r = 100$ is better for 3–6 tugs. Finally, only f_2 (all configurations except $f_2(1, 100)$) is able to steadily improve its normalised performance with an increasing number of tugs, reaching an improvement of 65–69 % compared to the static strategy for any configuration with a tug fleet of six tugs.

5 Discussion

In this paper we define the tug fleet optimisation problem as a 1D model and present a receding horizon genetic algorithm for solving this dynamic optimisation problem. By minimising a cost function, the algorithm determines positional trajectories for tugs to follow along the northern Norwegian coast such that the risk of oil tanker drifting accidents is reduced. We provide a set of potential cost functions with which the RHGA can be configured, effectively yielding a set of different TFO algorithms that each can be examined with respect to dynamic tug fleet positioning. In order to compare the merit of such TFO algorithms (or configurations of the RHGA), we propose two evaluation heuristics and test them by means of a computational simulation study.

5.1 Assessment of the Evaluation Heuristics

Both evaluation heuristics are able to quantify the performance of TFO algorithms designed to solve the TFO problem as defined in this paper. The small standard error for both heuristics is obtained by employing a large number of simulation scenarios (total 1,600) and means that the uncertainty in the means of both h_1 and h_2 is small. This is important in order to reliably measure the performance of TFO algorithms.

The general effect of increasing the number of oil tankers is that a static strategy will become increasingly suitable, whereas a dynamic scheme such as the RHGA configurations tested here will become less important. Thus, it is very impressive that cost function f_2 is able to increase its performance relative to the static strategy as measured by both evaluation heuristics, even with five or six tugs, and for all its configurations. The evaluation heuristics show that the RHGA configured with cost function f_2 has the best overall performance, with most configurations outperforming the other cost functions. The best choice of safe region for f_2 was $r = 50$, whereas the best power setting was $e = 1$ for h_1

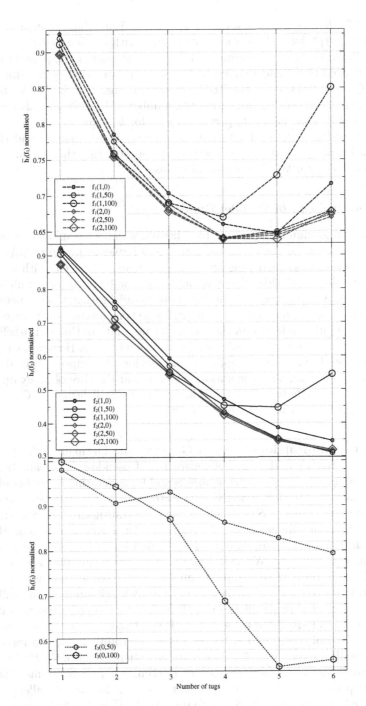

Fig. 6. Mean h_2 for cost functions f_1 (top), f_2 (middle), and f_3 (bottom), normalised by the static strategy.

and $e = 2$ for h_2, for 1–3 or 1–4 tugs, respectively. With these configurations, f_2 was able to outperform the static strategy even with one less tug.

Cost function f_3 also performs well if $r = 100$, and is comparable with f_2 when evaluated by h_1 and to a lesser extent when evaluated by h_2. However, if $r = 50$, f_3 is the worst of all the RHGA configurations. The similarities between cost function f_3 and h_1 in measuring the number of tankers are salvageable or not is probably what makes f_3 perform better for h_1 than for h_2.

Cost function f_1 is similar to but consistently worse than f_2 and the best configuration of f_3 and should be rejected. We propose that this is a direct result of its flaw that we have documented previously [5].

5.2   The Problem of Parallel Randomness

Functional programming languages like Haskell, which we employ here, are considered well suited for the clean and structured implementation of GAs. Moreover, the purely-functional programming paradigm of Haskell, with no global state or functions with side-effects, is particularly well suited for parallelisation. In addition, GAs lend themselves naturally to parallelisation, since genetic operations such as mating and mutation are performed on chromosome pairs or single chromosomes independent from the rest of the population. Hence, Haskell seems like a very good choice for parallel implementation of GAs. However, these very properties listed above that simplify parallelisation also impose strict demands on pseudo-random number generation, which lies at the core of many optimisation algorithms, including GAs. The challenges of parallelising the GA and the PRNG in Haskell has been discussed in [20].

Parallel Pseudo-Random Number Generation. The main challenge in parallelising the RHGA is the parallel generation of pseudo-random numbers that are required for the GA to evolve solutions and for generating a number of simulation scenarios for testing. A PRNG is a state machine that can be expressed as a function $\text{next} : S \to R \times S$ that defines the state transitions with S as the state space and R as the sample space. We may consider a generator g as an element $g \in S$ such that the function $\text{next}\, g = (r, g')$ provides a new generator g' as well as the random number r. Without a global state, managing the generator state is a challenge in pure functional programming. In the sequential case, monads is a good way to overcome the challenge, but parallelisation remains a challenge.

A naïve solution to parallelise the PRNG, is to use concurrency, with one thread dedicated to maintaining a global PRNG. However, this has several drawbacks. First of all, the program becomes non-deterministic, due to race conditions. Secondly, heavy parallelism can make a bottleneck out of the PRNG. With explicit multi-threading, as in concurrent Haskell, it is possible to maintain separate, independent PRNGs for each thread. However, for data-parallel problems such as the GA, it is generally better to use parallel Haskell which leaves more of the parallelisation work for the compiler, and where the threads are not visible to the programmer. This requires splittable PRNGs.

Splittable Pseudo-Random Number Generators. Splittable PRNGs is a concept originally introduced by [21] and has been adopted by the standard library of Haskell. In addition to the aforementioned next-function one has a function split : $\mathcal{S} \to \mathcal{S} \times \mathcal{S}$, which is very flexible and accommodates different forms of parallelisation. For example, whenever a function requires the PRNG, the caller can split its generator state g, to get $(g_1, g_2) = \text{split } g$, and pass g_1 to the callee and retain g_2 for further probabilistic processing. No generator has to be returned from the function, and thus the PRNG does not create any computational dependency. Thus the compiler is free to parallelise the function call and subsequent expressions.

Although the split-functionality we describe sounds like an ideal mechanism for parallel randomness, it has recently been shown that the standard library implementation in Haskell is unsound [20, 22]. Specifically, when performing repeated split-operations in a Monte Carlo tree fashion, the corresponding pseudo-random sequences fail to be uncorrelated as required, with high risk of dependencies within a small neighbourhood in the Monte Carlo tree.

Splittable PRNGs as an idea have been studied since the 1980s [23], but even so, they have received relatively little attention in the literature. The only trustworthy constructions we have found are very recent, from 2013 [22] and 2014 [24], respectively. A detailed assessment of known constructions is currently under preparation [25].

5.3   Future Work

The main hurdle before our RHGA can be used in real-world systems is to test and verify it under realistic conditions. This includes considering historical data of oil tanker traffic, realistic estimates of the variable maximum tug speeds attainable under various conditions, realistic drift trajectories and cross point distributions, downtime of tugs due to secondary missions or change of crew, and so on. It may also be necessary to extend the algorithm to 2D, in particular high risk scenarios where oil tankers enter or leave port and therefore are much closer to land than when sailing along the TSS corridor. In a manuscript soon to be submitted to a journal [26], the first author and colleagues address a number of these considerations by means of a 2D nonlinear mixed integer programming (MIP) formulation that we will discuss below.

Nonlinear Mixed Integer Programming for Tug Fleet Optimisation. [26] have recently extended earlier work [12] that used receding horizon control in combination with MIP instead of a GA to solve the dynamic TFO problem. In their nonlinear MIP formulation, the coastal region of interest is extended from 1D by dividing it into a finite number of cells yielding a 2D discretisation. To reduce the computational cost, the nonlinear MIP model is linearised by two different methods, MIP-U and MIP-L, that provide upper and lower bounds, respectively, on the solution. Based on recent work on theoretical drift prediction [27], realistic potential drift trajectories are estimated and used in the computational experiments. Adopting the findings of [28,29], and others, Assimizele and colleagues attempt to use minimisation of clean-up costs, socioeconomic losses and environmental costs associated with oil spill from grounding accidents as optimisation objectives. Finally, computational results are obtained by incorporating real historical ship traffic data and near grounding incidents provided us by the NCA, thus allowing for a comparison between real historical movements made by the fleet of tugs and the hypothetical trajectories suggested by the solution to the nonlinear MIP formulation.

5.4   Final Remarks

Are we able to say that TFO algorithms can make better choices in particular situations than a human operator at a VTS centre? A positive and justified answer to this question is needed before the NCA or others will be willing to invest in developing a software prototype of the kind of TFO algorithms we present here. To justify such a claim, sound risk assessment based on accurate models of potential drift trajectories that incorporate real-time and forecast weather and ocean current data, as well as real historic traffic data, is necessary. Although the problem is challenging, we do believe it is solvable to a satisfactory degree and we do welcome the prospect of TFO algorithms being adopted as decision-support tools for VTS centres around the world.

Acknowledgements. The SoftICE lab is grateful for the support provided by Regionalt Forskningsfond Midt-Norge and the Research Council of Norway through the project *Dynamic Resource Allocation with Maritime Application* (DRAMA), grant no. ES504913.

References

1. Bye, R.T., Schaathun, H.G.: Evaluation heuristics for tug fleet optimisation algorithms: a computational simulation study of a receding horizon genetic algorithm. In: Proceedings of the 4th International Conference on Operations Research and Enterprise Systems (ICORES 2015), pp. 270–282 (2015)
2. Havforskningsinstituttet: Fisken og havet, særnummer 1a–2010: Det faglige grunnlaget for oppdateringen av forvaltningsplanen for Barentshavet og havområdene utenfor Lofoten. Technical report, Institute of Marine Research (Havforskningsinstituttet) (2010)
3. Bye, R.T., van Albada, S.B., Yndestad, H.: A receding horizon genetic algorithm for dynamic multi-target assignment and tracking: a case study on the optimal positioning of tug vessels along the northern Norwegian coast. In: Proceedings of the International Conference on Evolutionary Computation (ICEC 2010) – part of the International Joint Conference on Computational Intelligence (IJCCI 2010), pp. 114–125 (2010)
4. Bye, R.T.: A receding horizon genetic algorithm for dynamic resource allocation: a case study on optimal positioning of tugs. In: Madani, K., Dourado Correia, A., Rosa, A., Filipe, J. (eds.) Computational Intelligence. SCI, vol. 399, pp. 131–148. Springer, Heidelberg (2012)

5. Bye, R.T., Schaathun, H.G.: An improved receding horizon genetic algorithm for the tug fleet optimisation problem. In: Proceedings of the 28th European Conference on Modelling and Simulation, pp. 682–690 (2014)
6. Vardø VTS: The Vardø Vessel Traffic Service - For increased safety at sea. Information pamphlet (2011). http://kystverket.no
7. Vardø, VTS: Annual report on petroleum transport statistics. Technical report, Norwegian Coastal Administration (2014)
8. Vardø, VTS: Annual report on petroleum incidents. Technical report, Norwegian Coastal Administration (2015)
9. Vardø, VTS: Annual report on petroleum incidents. Technical report, Norwegian Coastal Administration (2014)
10. Eide, M.S., Endresen, Ø., Breivik, Ø., Brude, O.W., Ellingsen, I.H., Røang, K., Hauge, J., Brett, P.O.: Prevention of oil spill from shipping by modelling of dynamic risk. Mar. Pollut. Bull. **54**, 1619–1633 (2007)
11. Eide, M.S., Endresen, Ø., Brett, P.O., Ervik, J.L., Røang, K.: Intelligent ship traffic monitoring for oil spill prevention: risk based decision support building on AIS. Mar. Pollut. Bull. **54**, 145–148 (2007)
12. Assimizele, B., Oppen, J., Bye, R.T.: A sustainable model for optimal dynamic allocation of patrol tugs to oil tankers. In: Proceedings of the 27th European Conference on Modelling and Simulation, pp. 801–807 (2013)
13. Holland, J.H.: Adaptation in Natural and Artificial Systems: An Introductory Analysis with Applications to Biology, Control, and Artificial Intelligence. University of Michigan Press, Oxford (1975)
14. Goldberg, D.E.: Genetic Algorithms in Search, Optimization, and Machine Learning. Addison-Wesley Professional, Reading (1989)
15. Haupt, R.L., Haupt, S.E.: Practical Genetic Algorithms, 2nd edn. Wiley, New York (2004)
16. Goodwin, G.C., Graebe, S.F., Salgado, M.E.: Control System Design. Prentice Hall, Upper Saddle River (2001)
17. Maciejowski, J.M.: Predictive Control with Constraints, 1st edn. Prentice Hall, Upper Saddle River (2002)
18. Rossiter, J.A.: Model-Based Predictive Control. CRC Press, Boca Raton (2004)
19. Det Norske Veritas: Rapport Nr. 2009–1016. Revisjon Nr. 01. Tiltaksanalyse – Fartsgrenser for skip som opererer i norske farvann. Technical report, Sjøfartsdirektoratet (2009)
20. Schaathun, H.G.: Parallell slump (Om å parallellisera genetiske algoritmar i Haskell). In: Norsk Informatikkonferanse (NIK 2014) (2014)
21. Burton, F.W., Page, R.L.: Distributed random number generation. J. Funct. Program. **2**, 203–212 (1992)
22. Claessen, K., Pałka, M.H.: Splittable pseudorandom number generators using cryptographic hashing. ACM SIGPLAN Not. **48**, 47–58 (2013). ACM
23. Frederickson, P., Hiromoto, R., Jordan, T.L., Smith, B., Warnock, T.: Pseudorandom trees in Monte Carlo. Parallel Comput. **1**, 175–180 (1984)
24. Steele Jr., G.L., Lea, D., Flood, C.H.: Fast splittable pseudorandom number generators. ACM SIGPLAN Not. **49**, 453–472 (2014)
25. Schaathun, H.G.: Evaluation of splittable pseudo-random generators. J. Funct. Program. (2015). Manuscript under revision
26. Assimizele, B., Royset, J.O., Bye, R.T., Oppen, J.: Preventing Environmental Disasters from Grounding Accidents: A Case Study of Tugboat Positioning along the Norwegian Coast (2015, submitted)

27. Ni, Z., Qiu, Z., Su, T.: On predicting boat drift for search and rescue. Ocean Eng. **37**, 1169–1179 (2010)
28. Vanem, E., Endresen, Ø., Skjong, R.: Cost-effectiveness criteria for marine oil spill preventive measures. Reliab. Eng. Syst. Saf. **93**, 1354–1368 (2008)
29. Kontovas, C.A., Psaraftis, H.N., Ventikos, N.P.: An empirical analysis of IOPCF oil spill cost data. Mar. Pollut. Bull. **60**, 1455–1466 (2010)

Extended Decomposition for Mixed Integer Programming to Solve a Workforce Scheduling and Routing Problem

Wasakorn Laesanklang[✉], Rodrigo Lankaites Pinheiro,
Haneen Algethami, and Dario Landa-Silva

ASAP Research Group, School of Computer Science,
The University of Nottingham, Jubilee Campus, Wollaton Road,
Nottingham NG8 1BB, UK
{psxwl3,rodrigo.lankaitespinheiro,psxha7,
dario.landasilva}@nottingham.ac.uk

Abstract. We propose an approach based on mixed integer programming (MIP) with decomposition to solve a workforce scheduling and routing problem, in which a set of workers should be assigned to tasks that are distributed across different geographical locations. We present a mixed integer programming model that incorporates important real-world features of the problem such as defined geographical regions and flexibility in the workers' availability. We decompose the problem based on geographical areas. The quality of the overall solution is affected by the ordering in which the sub-problems are tackled. Hence, we investigate different ordering strategies to solve the sub-problems. We also use a procedure to have additional workforce from neighbouring regions and this helps to improve results in some instances. We also developed a genetic algorithm to compare the results produced by the decomposition methods. Our experimental results show that although the decomposition method does not always outperform the genetic algorithm, it finds high quality solutions in practical computational times using an exact optimization method.

Keywords: Workforce scheduling · Routing problem · Mixed integer programming · Problem decomposition · Genetic algorithm

1 Introduction

The workforce scheduling and routing problem (WSRP) is a difficult problem that arises in industries like home care, health care, patrol service, meter reading, etc. [9]. One specific example of WSRP is home health care services where nurses or care workers should deliver care services to patients at their home. Solving the problem involves producing a job schedule and a route for each worker while satisfying the business requirements and considering workers qualifications and skills, task requirements, travelling distances, etc. It is usually expected that the solution gives the lowest operational cost.

© Springer International Publishing Switzerland 2015
D. de Werra et al. (Eds.): ICORES 2015, CCIS 577, pp. 191–211, 2015.
DOI: 10.1007/978-3-319-27680-9_12

Developing automated solution methods to solve WSRP scenarios is a current research challenge as reflected by recent published research [17,22,23]. Solving an instance of WSRP often requires the expertise and knowledge of a human planner [4]. In this research, we are working with an industrial partner who provides scheduling services for businesses and other organisations facing this type of problems. The data sets considered here come from real-world scenarios. A particular feature is that 'working areas' or 'regions' are pre-defined and this affects the difficulty of the problem being tackled.

Tackling WSRP with exact optimization methods to produce solutions in practice is still a research challenge. Since obtaining an optimal schedule is the ultimate goal, exact methods like mathematical programming are a suitable approach. However, proven optimality with exact methods has been shown mainly on problem instances of limited size [7,8]. Problem instances faced in practice are larger and for solving them, heuristic methods are usually considered more efficient in terms of computation time [2,27]. The motivation for our work is to develop a solution approach based on exact optimization to tackle real-world WSRP instances.

In this paper, we propose a decomposition approach that uses mixed integer programming (MIP) to tackle WSRP instances of practical size. The proposed method splits the problem into sub-problems according to defined geographical areas. Our computational experiments show that the approach allows to explore the trade-off between computation efficiency and solution quality. We also use a process that brings additional workforce to understaffed regions from neighbouring regions. This process is applied when required before further splitting the regions into smaller sub-problems. The results from our experiments suggest that the success of the proposed decomposition varies according to the problem instance, which provides us with ideas for future research.

The main contribution of this paper is to show that the ordering in which sub-problems in workforce scheduling and routing are tackled within a decomposition approach, has an effect on the computational efficiency and achieved solution quality. Following this, some ordering strategies are proposed to achieve solutions of good quality in practical computation time. Moreover, we also compare the results produced with the decomposition method to the results obtained with a standard genetic algorithm and to the optimal solution when available.

Section 2 reviews related works in the literature and establishes the motivation for the research in this paper. Section 3 gives the problem definition and formulation for the WSRP considered here. Section 4 describes the proposed MIP with decomposition approach and the experimental study, including description of test data instances. The final section summarises the paper and outlines some of the proposed future work.

2   Literature Review

Solving integer programming formulations for larger problem instances still has its limitations in terms of computation time. Mathematical programming has been used in the literature to tackle some WSRP. Examples include linear programming [3], integer programming [20] and mixed integer programming [7,8,13,30]. To solve real-world sized problems, works in the literature often resource to heuristic or hybrid algorithms [2,6,14]. There are some improved exact methods, like branch and price [8,13,30], that can deal with large scenarios. Branch and price requires problem reformulation which usually involves Dantzig-Wolfe decomposition to compute a tighter relaxation bound [15,33]. The algorithm also requires two steps to repeatedly solve the problem in order to improve the solution.

Decomposition techniques are another good alternative to apply exact optimization methods to large integer programming formulations. The basic idea is to transform or split the problem into smaller sub-problems. This technique has been applied in various problem domains. For example, Benders' decomposition was used to produce solutions for large instances of the aircraft routing and crew scheduling problem [11,24]. Benders' decomposition is suitable for problems with exclusive sub-problem sets or problems that show some block structures linked by constraints [5]. In another example of decomposition, [26] split the warehouse location-routing problem into three smaller problems: the complete multi-depot vehicle-dispatch problem, the warehouse location-allocation problem and the multi-depot routing-allocation problem. These three smaller problems were solved in phases and each of them was formulated with mathematical programming and solved by an exact solver. For detailed reviews of decomposition approaches see [29,34].

Decomposition techniques have also been applied within heuristic approaches using some form of clustering. For example, [31] tackled a large vehicle routing problem by decomposing it into sub-problems. Each sub-problem was a cluster of customers assigned to a vehicle which then became a travelling salesman problem. The sub-problem size is controlled by splitting a large sub-problem to shrink the corresponding cluster. Similar ideas were applied in a hybrid heuristic for generating multi-carrier transportation plans [21].

In this work we propose a decomposition approach that uses mixed integer programming (MIP) to tackle workforce scheduling and routing problem instances arising in real-world scenarios. For this, we also present an MIP formulation that incorporates features of the WSRP scenarios faced by our industrial partner. The proposed decomposition technique does not require some formulation structure like in Benders' decomposition neither uses a heuristic solver. Our approach harness the power of exact optimization solvers while decomposing the problem instances in a way that is meaningful to practice.

3   Problem Description and Formulation

The goal in WSRP is to assign each worker to undertake a set of tasks across a set of geographical locations. A path is the series of tasks to be performed by a worker within the planning period. A good quality solution consists of a set of shortest paths representing the task assignments for each worker at the lowest cost. The solution should also respect other conditions such as task pre-requisites, required travelling time between locations, defined appointment times, workers' skills, workers availability, restricted working regions, working time limits, etc.

Consider a graph $G = (V, E)$ where $V = T \cup D \cup D'$ represents the union of a set of tasks (each task as a location) T, a set of start locations D and a set of end locations D' while E represents a set of links between two locations (e.g. between two task locations or between the worker's home and a task location). The set of workers is denoted by C. Binary decision variable $x_{i,j}^c = 1$ if worker $c \in C$ is assigned to a task $j \in T$ after finishing task $i \in T$, $x_{i,j}^c = 0$ otherwise. Note that elements of T are referred here as tasks but also each task has an associated location.

In real-world scenarios like the ones considered here, the available skilled workforce is often not sufficient to cover all the required tasks. We introduce a dummy worker to address this issue (through an integer decision variable y_j) that takes any uncovered task that cannot be assigned to the available workforce [1,30]. In our problem some tasks may require more than one worker. We denote the number of workers required to perform a task by b_j. Therefore, the required number of workers must be assigned to each task. The dummy worker can be used for that effect too, even taking the whole task as $y_j = b_j$. Then, the assignment of tasks is represented by (1).

$$\sum_{c \in C} \sum_{i \in D \cup T} x_{i,j}^c + y_j = b_j \qquad , \forall j \in T \tag{1}$$

The sequence of tasks performed by a worker is represented as a path for visiting task locations, hence the number of workers arriving at one location must be equal to the number of workers leaving that task location so that either workers are assigned to the next task or go home. Then, the path constraint is represented by (2).

$$\sum_{i \in D \cup T} x_{i,j}^c = \sum_{k \in D' \cup T} x_{j,k}^c \quad , \forall j \in T, \forall c \in C \tag{2}$$

Workers must start and end their paths from their specific location (e.g. their home or a central office) as given by (3) and (4). Since D and D' are sets of start and end locations respectively, these two constraints indicate the start and end locations for each worker. Also, workers leave their start location and enter their end location at most once (although the start and end locations can be different) as expressed by (5) and (6) respectively. Note that a worker does not leave his start location if he is not assigned to work. This is different from the common case in the literature where all workers leave their start location. In our problem instances, the specific start and end locations are provided for every worker.

$$\sum_{j \in D' \cup T} x_{k,j}^c \geq \sum_{j \in D' \cup T} x_{i,j}^c \quad , \forall c \in C, \forall i \in T, \exists k \in D \tag{3}$$

$$\sum_{i \in D \cup T} x_{i,k}^c \geq \sum_{i \in D \cup T} x_{i,j}^c \quad , \forall c \in C, \forall j \in T, \exists k \in D' \tag{4}$$

$$\sum_{j \in D' \cup T} x_{i,j}^c \leq 1 \qquad , \forall i \in D, \forall c \in C \tag{5}$$

$$\sum_{i \in D \cup T} x_{i,j}^c \leq 1 \qquad , \forall j \in D', \forall c \in C \tag{6}$$

Let S be the set of skills and $s \in S$ a particular skill. For worker c the qualification level on skill s is q_s^c and for task j the requirement of skill s is $r_{s,j}$. Hence, worker c can be assigned to task j only if the worker has the required qualifications level for skill s, that is, $q_s^c \geq r_{s,j}$. Then, in our model the multi-skill qualification requirements are represented by (7).

$$x_{i,j}^c r_{s,j} \leq q_s^c \qquad , \forall c \in C, \forall i \in D \cup T, \forall j \in T, \forall s \in S \tag{7}$$

Also, travel time between task locations must be feasible. Decision variable a_i^c takes a positive fractional value that gives the worker arrival time to the location of task i. Note that the maximum arrival time value is 1440 which is equivalent to the 24th hour of the day. Let a_i^c, a_j^c be the arrival times of worker c to the locations of task i and task j respectively. Let $t_{i,j}$ be the travelling time between the locations of tasks i and j. Let δ_i be the duration of task i. Then, if worker c is assigned to perform task j after completing task i, inequality (8) (M is a large constant number) expresses the arrival on time requirement.

$$a_j^c + M(1 - x_{i,j}^c) \geq a_i^c + x_{i,j}^c t_{i,j} + \delta_i \qquad , \forall c \in C, \forall i \in D \cup T, \forall j \in D' \cup T \tag{8}$$

An arrival time window is also defined for task j and the worker should not arrive earlier than w_j^L or later than w_j^U, as expressed by (9).

$$w_j^L \leq a_j^c \leq w_j^U \qquad , \forall j \in T, \forall c \in C \tag{9}$$

An important feature of our WSRP scenarios is that working regulations and availability can be specific for each worker. In the problems considered here, this refers to long breaks between shifts (short breaks within the working shift are not considered), days-off, working shift duration, maximum working hours, and specific worker preferences (e.g. late morning, afternoon only, whole day, overnight). We adopt a flexible availability constraint from an optimization of daily scheduling for home health care services [32]. Any task assignment at time a_j^c including the task duration δ_j should lie in between the shift starting time α_L^c and the shift ending time α_U^c. The availability parameters α_L^c and α_U^c are real numbers defined for each worker c. A task assigned outside the shift is charged as additional expense, hence binary decision variable $\omega_j = 1$ if this is the case and $\omega_j = 0$ otherwise. Then, individual availability constraints are denoted by (10) and (11) while the working hours limit (h^c) constraint is denoted by (12).

$$\alpha_L^c - a_j^c \leq M(1 - x_{i,j}^c + \omega_j) \qquad , \forall c \in C, \forall i \in D \cup T, \forall j \in T \tag{10}$$

$$a_j^c + \delta_j - \alpha_U^c \leq M(1 - x_{i,j}^c + \omega_j) \qquad , \forall c \in C, \forall i \in D \cup T, \forall j \in T \tag{11}$$

$$\sum_{i \in D \cup T} \sum_{j \in T} x_{i,j}^c \delta_j \leq h^c \qquad , \forall c \in C \tag{12}$$

Another important feature of our WSRP scenarios is that workers have preferred geographical areas for working but the decision maker can still request workers to work outside those preferred regions. We formulate this in (13) where binary parameter $\gamma_j^c = 1$ if worker c is willing and able to work at the location of task j, $\gamma_j^c = 0$ otherwise, and binary decision variable $\psi_j = 1$ if worker c is forced to work outside their defined regions.

$$\sum_{i \in D \cup T} x_{i,j}^c - \psi_j \leq \gamma_j^c \quad , \forall c \in C, \forall j \in T \tag{13}$$

Most of the above constraint formulations exist in literature but not all. Common constraints (see also [9]) such as path constraint (2), skill and qualification (7) and time windows (9) form the basic structure of the scheduling and routing problem [1,10,12,13,30]. Tailor cut constraints adopted from literature are the availability constraints (10,11) while the constraints that required further adaptation to our problem features are the working region (13) (implemented as soft constraint) and start-end paths (3–6).

Table 1 summarises the constraints in our MIP model. Given our real-world data sets, some are implemented as soft constraints. For example, workers can be forced to work outside their predefined regions and availability. Also, tasks can be left unassigned (assigned to the dummy worker). These features are quite important to maintain the practical applicability of our model and solution approach.

Table 1. Summary of problem requirements and type of constraints.

	Hard	Soft		Hard	Soft
Job assignment (Cons. (1))		*	Start-end paths (Cons. (3)–(6))	*	
Path constraint (Cons. (2))	*		Travel time feasibility (Cons. (8))	*	
Time windows (Cons. (9))	*		Skill and qualification (Cons. (7))	*	
Working hours (Cons. (12))	*		Worker availability (Cons.(10), (11))	a	*
Working regions(Cons. (13))		*			

[a]Hard constraints (15) and (16) are described in Sect. 4.2 and only apply to the decomposition model.

The objective function (14) involves three costs: monetary cost, preferences penalty cost and soft constraints penalty cost.

$$\text{Min} \sum_{c \in C} \sum_{i \in D \cup T} \sum_{j \in D' \cup T} \lambda_1 \left(d_{i,j} + p_j^c \right) x_{i,j}^c + \sum_{c \in C} \sum_{i \in D \cup T} \sum_{j \in D' \cup T} \lambda_2 \rho_j^c x_{i,j}^c$$
$$+ \sum_{j \in T} (\lambda_3(\omega_j + \psi_j) + \lambda_4 y_j) \tag{14}$$

The first term in (14) is the monetary cost and includes the travelling cost $d_{i,j}$ of going from location of task i to the location of task j, and the payment p_j^c for worker c to perform task j. The second term in (14) is the preference penalty cost denoted by $\rho_j^c \geq 0$ and is a summation of penalties for not meeting worker-client preferences, required skill preferences and working region preferences. This penalty value can go from 0 to 2 and $\rho_j^c = 0$ when all preferences are met, while this penalty value grows higher as the preference level of assigning worker c to task j decreases. The third term in (14) is the soft constraints penalty cost due to the violation of the three soft constraints in the model. The job assignment constraint has the highest priority, so a violation of this constraint type costs more than a violation of the other two constraints. The worker availability and working regions constraints have the same priority. Note that the working regions constraint is involved in two costs. If the worker is assigned a task in a non-preferred region then this is a constraint penalty cost. If the worker is assigned to one of the preferred regions this is quantified as a preference penalty cost according to the degree in which the region is preferred by the worker (several working regions with different preference levels). Note that $\lambda_1, \lambda_2, \lambda_3$ and λ_4 are weights that set the priorities between objectives as $\lambda_1 < \lambda_2 < \lambda_3 < \lambda_4$. The value assigned to these weights are included with the data instances.

The above MIP model corresponds to the integrated scheduling and routing problem. Solving this model with an exact optimization method is not practical considering our real-world problem instances. Hence, we apply a decomposition technique.

4 Decomposition Approach and Study

In order to reduce the overall computational time for solving real-world instances of the integrated workforce scheduling and routing problem, we now present a decomposition method. First, we describe the features of our problem instances as this will help to explain the proposed decomposition approach. Later, the method is described and experimental results are provided.

4.1 Test Instances

For the present work, we prepared some test instances using real-world data corresponding to home care scenarios in the UK, provided by our industrial partner. A problem instance P has a set of nodes V. Recall from Sect. 3 that $V = D \cup T \cup D'$. Also, some of the tasks $\{j_1, j_2, \ldots, j_n\}$ in T share the same geographical location $\kappa \in K$, where K is a set of geographical locations. A group of locations are assembled as a geographical region or working region $a \in A$. Note that $a \subseteq K$ and A is a partition. Also, an individual worker c may work on one or several geographical regions. As noted above, a key aspect of our scenarios is that several tasks might be required at one particular location. Each individual task may have different required skills, worker preferences and worker cost.

We took four real-world scenarios and prepared a data set from each. Although the instances in each data set come from the same scenario, each instance is formed from a different planning time giving a variation in the available human resources and task requirements. In our data, the start and end

Table 2. The test data sets.

Instance	$|C|$	$|K|$	$|T|$	$|A|$	Instance	$|C|$	$|K|$	$|T|$	$|A|$
WSRP-A-01	23	25	31	6	WSRP-B-01	25	27	36	6
WSRP-A-02	22	24	31	4	WSRP-B-02	25	11	12	4
WSRP-A-03	22	28	38	5	WSRP-B-03	34	43	69	6
WSRP-A-04	19	22	28	3	WSRP-B-04	34	14	30	4
WSRP-A-05	19	9	13	3	WSRP-B-05	32	38	61	8
WSRP-A-06	21	22	28	7	WSRP-B-06	32	38	57	7
WSRP-A-07	21	9	13	3	WSRP-B-07	32	38	61	7
WSRP-D-01	164	233	483	13	WSRP-F-01	805	477	1211	45
WSRP-D-02	166	215	454	12	WSRP-F-02	769	496	1243	46
WSRP-D-03	174	279	585	15	WSRP-F-03	898	582	1479	54
WSRP-D-04	174	237	520	15	WSRP-F-04	789	513	1448	47
WSRP-D-05	173	259	538	15	WSRP-F-05	883	626	1599	59
WSRP-D-06	174	291	610	15	WSRP-F-06	783	565	1582	44
WSRP-D-07	173	293	611	15	WSRP-F-07	1011	711	1726	64

$|C|$ = number of workers.
$|K|$ = number of task locations.
$|T|$ = number of required tasks.
$|A|$ = number of working regions.

locations of a worker are the same ($d = d'$). Table 2 shows the main features of the test instances: the number of available workers $|C|$, the number of task locations $|K|$, the numbers of tasks $|T|$ and the number of predefined geographical regions $|A|$. In terms of size, instances WSRP-A-(01-07) and WSRP-B-(01-07) are considered small with around 19–34 workers and 13–69 tasks. The optimal solution for each of these instances can be found in less than 5 min. Instances WSRP-D-(01-07) and WSRP-F-(01-07) are considered large with more than 100 workers and 400 tasks. These large instances cannot yet be solved to optimality in practical computation time. In our experimental study, we use the small instances to validate the proposed decomposition approach as we can compare to the optimal solutions. Moreover, the experimental results show the suitability of the decomposition approach in tackling the large instances using an exact optimization solver.

4.2 Geographical Decomposition with Conflict Avoidance

In this paper, the workforce scheduling and routing problem is decomposed into working regions as this is a key feature of the scenarios provided by our industrial partner. Since we decompose the problem into sub-problems to deal with the larger size more efficiently, by solving the sub-problems one at a time in a given sequence, we can no longer guarantee overall optimality.

Algorithm 1. Geographical decomposition with conflict avoidance (GDCA).

Data: Problem instance $P = \{C, A\}$

1 initialization: For worker $c \in C$, define earliest availability vector $\beta_L = (\beta_L^c)$ and latest availability vector $\beta_U = (\beta_U^c)$;

2 Split problem P by regions denoted as $P_i = \{C, a\}, a \in A, i = 1...|A|$;

3 **forall the** $P_i \in P$ **do**

4   Solve P_i with availability α_L and α_U by CPLEX solver$\rightarrow \Phi_i$;

5   Update availability vector β_L and β_U;

6 **end**

7 Combine sub-problem solutions;

Basically, the decomposition method generates a sub-problem for each working region and solves each sub-problem in sequence. Worker assignment conflicts (i.e. a worker being assigned to different task locations at the same time) are avoided because each sub-problem is solved using only the reduced available workforce after solving the previous sub-problem.

Algorithm 1 presents the proposed geographical decomposition with conflict avoidance approach (GDCA). A problem instance P is split into several sub-problems P_i (step 2). A sub-problem P_i corresponds to a geographical region or working region $a \in A$. Some regions may generate a sub-problem that is too large. Hence, we further split them until the sub-problem has no more than around ten locations. Then, the sub-problems are solved in a given sequence (steps 3–6) and different solving sequences can lead to different solution quality. This is because the first sub-problem has access to the most workforce resources but subsequent sub-problems will have access to limited available workforce. Since worker assignment conflicts are avoided, this means that the hard constraints expressed by Eqs. (15) and (16) are enforced in this algorithm.

$$a_j^c + \delta_j - \beta_L^c \leq M(2 - x_{i,j}^c - \zeta^c) \quad , \forall c \in C, \forall i \in D \cup T, \forall j \in T \cup D' \quad (15)$$

$$\beta_U^c - a_j^c \leq M(1 - x_{i,j}^c + \zeta^c) \quad , \forall c \in C, \forall i \in D \cup T, \forall j \in T \cup D' \quad (16)$$

Here, β_L^c denotes the start of unavailable time and β_U^c denotes the end of unavailable time for worker c. Since the original model generates a continuous path for a worker, a path created under hard availability conditions is allocated either before or after the unavailability period. That is, a path which overlaps with the unavailability period defined by β_L^c and β_U^c is not allowed. The control variable ζ^c is applied for selecting only one side of the availability period. When $\zeta^c = 1$ the time interval before β_L^c is selected and if the $\zeta^c = 0$ the time interval after β_U^c is selected.

In our data, we know that the start location d and end location d' for a worker are the same. Therefore, we designed a sub-problem solutions combination process based on this assumption. During the **Combine sub-problem solutions** process (step 7), sub-problem solutions are combined together by connecting the worker's paths from each sub-problem to get a long single path. After this process, a worker leaves his start location and arrives to his end location only once. Suppose that $\Phi_1 = \{(x_{d,t_1}^c, a_1^c), (x_{t_1,d'}^c, a_{d_1}^c)\}$ is a solution to sub-problem P_1 representing the assignment of worker c from start location d to work on task t_1 and returning to end location d' and $\Phi_2 = \{(x_{d,t_2}^c, a_2^c), (x_{t_2,d'}^c, a_{d_2}^c)\}$ is a solution to sub-problem P_2 representing the assignment of the same worker c from starting location d to work on task t_2 and returning to ending location d'. Assume without loss of generality that $a_1^c < a_{d_1}^c < a_2^c < a_{d_2}^c$. The combining process redirects the arriving assignment to end location d' to task t_2 which gives a global solution as $\Phi = \{(x_{d,t_1}^c, a_1^c), (x_{t_1,t_2}^c, a_2^c), (x_{t_2,d'}^c, a_{d_2}^c)\}$. It is possible than in other scenarios of the WSRP, the start location and end location for a worker are different, we leave this for future work as it is not a feature of the scenarios tackled at present.

4.3 Experimental Study of the Decomposition Method

We conduct an experimental study to gather insights into the performance of the proposed geographical decomposition method. The flow of the study is depicted in Fig. 1. The figure outlines the three parts of the experimental design. First, on the left-hand side of the figure, the **permutation study** refers to solving the sub-problems in different order given by all the different permutations of the geographical regions. However, trying all permutations is practical only in small problems. Therefore, finding an effective ordering pattern is the second part of the experiment, **observation step** in the figure. This second part solved each sub-problem using all available workforce, i.e. ignoring if some workers were assigned in previous sub-problems. The third part analysed the results from the observation step in order to define some strategies to tackle the sub-problems. Based on this **strategies study**, some solving strategies were conceived. Listed in the figure are these ordering strategies: Asc-task, Desc-task, Asc-w&u, etc. More details about these ordering strategies are provided when describing the **Observation step** below. Finally, the solutions produced with the different ordering strategies are compared to the solutions produced by the permutation study to evaluate the performance of these ordering strategies.

Permutation Study. Since the number of permutations grows exponentially with the number of geographical regions, we performed the permutation study using only the instances with $|A| = 3$ and $|A| = 4$ geographical regions. Figure 2 shows the relative gap obtained for the small instances that have 3 regions. Each sub-figure shows the results for one instance when solved using the different permutation orders of the 3 regions. Each bar shows the relative gap between the solution by the decomposition method and the overall optimal solution. The figure shows that the quality of the obtained solutions for the different permutations fluctuates considerably. Closer inspection reveals that in these instances the geographical regions are very close to each other and sometimes there is an overlap between them. The result also reveals that some permutations clearly give better results. For example, permutation "1-2-3" for instance WSRP-A-04, permutations "1-2-3" and "2-1-3" for instance WSRP-A-05 and permutations "1-3-2" for instance WSRP-A-07.

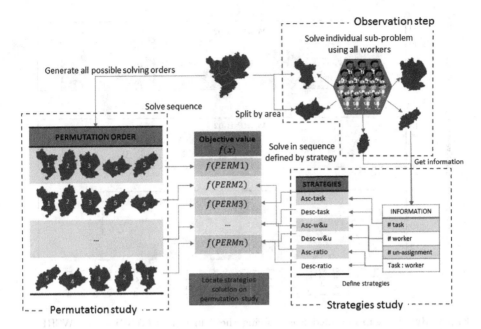

Fig. 1. Outline of the experimental study in three parts: permutation study, observation step and strategies study.

Figure 3 shows the relative gap obtained for the small instances that have 4 regions. Each sub-figure shows the result for one instance when solved using the permutation orders of the 4 regions. Each bar shows the relative gap between the solution by the decomposition method and the overall optimal solution. The figure reveals an interesting result from instance WSRP-B-02. The optimal solution value is obtained for every permutation. Closer inspection reveals that the decomposition method works very well on this instance because its geographical regions are well separated from each other. Therefore, the sub-problem solutions

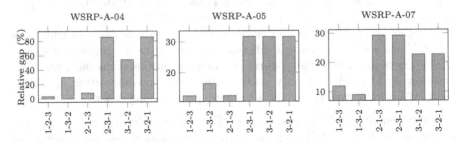

Fig. 2. Relative gap obtained from solving the 3 instances (WSRP-A-04, WSRP-A-05 and WSRP-A-07) with $|A| = 3$ using the different permutation orders. Each graph shows results for one instance. The bars represent the relative gap between the solution obtained with the decomposition method and the overall optimal solution.

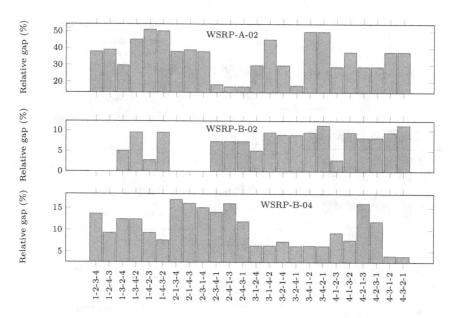

Fig. 3. Relative gap obtained from solving the 3 instances (WSRP-A-02, WSRP-B-02 and WSRP-B-04) with $|A| = 4$ using the different permutation orders. Each graph shows results for one instance. The bars represent the relative gap between the solution obtained with the decomposition method and the overall optimal solution.

are part of the complete overall solution and not many worker assignment conflicts arise when solving the sub-problems. For the other instances, WSRP-A-02 and WSRP-B-04, the quality of the obtained solutions fluctuates in the same way as in Fig. 2. Results in Fig. 3 indicate that some solutions obtained with the decomposition approach using some permutations have a considerable gap in quality compared to the overall optimal solution. The figure also shows that some permutations clearly give better results than others. For example, permutation "2-4-3-1" and "3-1-2-4" for instance WSRP-A-02, permutation "1-2-3-4", "1-2-4-3", "2-1-3-4", "2-1-4-3" and "2-3-1-4" for instance WSRP-B-02 and permutation "4-3-1-2" and "4-3-2-1" for instance WSRP-B-04.

The conclusion from this permutation study is that the order in which the sub-problems are solved matters differently according to the problem instance. More importantly, the results confirm our assumption that some particular permutation could produce a very good result in the decomposition approach. Hence, the next part of the study is to find a good solving order.

Observation Step. Here we solve each of the sub-problems using all available workers and collect the following values from the obtained solutions: number of tasks in the sub-problem (# task), minimum number of workers required in the solution (# min worker), number of unassigned tasks in the solution (# unassigned task) and the ratio of tasks to worker in the solution (task/worker ratio). Then, we defined six ordering strategies as follows. Increasing number of tasks

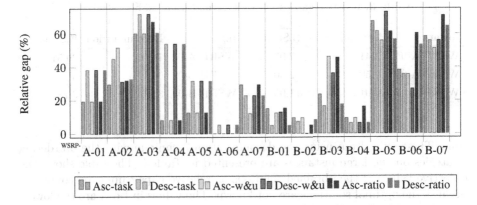

Fig. 4. Relative gap obtained from solving the 14 small instances using the 6 ordering strategies. Each bar for an instance represents the relative gap between a solution by the decomposition method using an ordering strategy and the overall optimal solution (Color figure online).

in the sub-problem (Asc-task); decreasing number of tasks in the sub-problem (Desc-task); increasing sum of minimum workers required and unassigned tasks (Asc-w&u); decreasing sum of minimum workers required and unassigned tasks (Desc-w&u); increasing ratio of tasks to worker (Asc-ratio) and decreasing ratio of tasks to worker (Desc-ratio).

Strategies Study. The GDCA approach is again executed using the 6 ordering strategies listed above to tackle the sub-problems in each problem instance. The results are presented in Fig. 4 which shows the relative gap for the 14 small instances in the WSRP-A and WSRP-B groups. Note that each bar represents the relative gap obtained with each strategy.

From Fig. 4, the decomposition technique with ordering strategies gives solutions with relative gaps up to 70 %. On average, the decomposition technique produces relative gap at 30.77 %. Moreover, we can see that some of the ordering strategies are more likely to produce better solutions than others. The best performing ordering strategy is Asc-w&u that gives 8 best solutions considering all 14 small instances. The average gap for the ordering strategies Asc-task, Desc-task, Asc-w&u, Desc-w&u, Asc-ratio and Desc-ratio are 27.14 %, 32.62 %, 27.39 %, 33.26 %, 31.61 % and 32.62 % respectively. Table 3 shows a comparison of relative gap between the best permutation order (see **Permutation study**) and the best ordering strategy. There are differences between the best strategies and the best permutation, the maximum being 12.56 % for instance WSRP-A-02. Two out of six solutions (instance WSRP-A-05 and WSRP-B-02) of the best ordering strategy match the solution from the best permutation. This shows that the ordering strategies are able to work well in other problem instances.

The decomposition method is also able to find solutions for the large instances whilst solving those problems as a whole is not practical in terms of computation

Table 3. Relative gap (%) of best permutation vs. best strategy.

Instance	B.Permutation	B.Strategy	Instance	B.Permutation	B.Strategy
WSRP-A-04	3.41	7.97	WSRP-A-02	16.95	29.51
WSRP-A-05	12.50	12.50	WSRP-B-02	0	0
WSRP-A-07	9.01	11.95	WSRP-B-04	3.91	6.28

time. The results from using the decomposition technique with the 6 ordering strategies on the large instances are presented in Table 4. The table shows the objective values of the obtained solutions as relative gaps cannot be computed because the optimal solutions are not known. The values in **bold** are the lowest cost (best objective value) obtained among the six strategies. The table shows that as a whole, Desc-task gives six best solutions, Desc-ratio gives four best solutions, Asc-w&u gives two best solutions, Desc-w&u and Asc-task give one best solution while the Asc-ratio gives no best solution. On average, the Desc-task strategy gives the lowest cost solution, around 15.42 % less than the highest average cost strategy (Asc-task).

Figure 5 shows, according to the problem size, the computation times used by the decomposition approach using the different ordering strategies and the time used to find the overall optimal solution. Each sub-figure presents the problem

Table 4. Objective value obtained from solving large instances using six ordering strategies.

Instance	Asc-task	Desc-task	Asc-w&u	Desc-w&u	Asc-ratio	Desc-ratio
D-01	1,688.07	496.45	1,549.04	765.48	1,301.04	**240.98**
D-02	860.50	**372.94**	496.47	495.44	984.98	732.97
D-03	2,624.84	3,213.32	2,619.33	3,836.84	**1,690.81**	3,839.34
D-04	312.43	418.89	303.45	**283.91**	314.42	420.41
D-05	408.42	243.89	1,113.47	253.91	401.45	**241.89**
D-06	**307.55**	1,411.27	945.60	1,582.52	634.05	1,729.29
D-07	1,112.80	753.28	**292.55**	604.01	293.53	1,077.28
F-01	73,286	64,305	71,430	72,040	75,760	**63,680**
F-02	81,852	**73,291**	76,460	80,569	86,906	74,860
F-03	141,060	115,235	140,258	120,715	148,092	**116,011**
F-04	111,671	102,994	105,262	109,411	113,557	**91,670**
F-05	127,476	**101,438**	113,403	105,284	112,995	103,156
F-06	105,595	**76,007**	88,702	84,050	107,281	84,050
F-07	199,160	**176,541**	194,525	178,387	218,058	178,387
Average	60,529.69	**51,194.39**	56,954.29	54,162.73	62,019.34	51,435.43

Bold text refers to the best solution.

instances classified by their size (number of items is $|T| + |C|$). Each line represents the time used by the ordering strategy in solving the group of 14 problem instances. As noted before, the time to find the optimal solution represented by [- - -] is available only for the small instances. For the smaller instances which smaller than instance B-06 (has 89 items), the computation time used by the decomposition method is not much different from the time used to find the optimal solution. The computation time used to find the optimal solution grows significantly for instances B-06 to B-03. Note that instance WSRP-B-03 which has 109 items uses 5,419 s for finding optimality. For the large instances, it is shown that the computation time used by the decomposition method starts from 17 min (1,060 s) to above 6 h (22,478 s). Also, for the large instances the average computation time used by six strategies are 4,620, 3,098, 7,451, 6,348, 7,640 and 7,048 s respectively. The average processing time shows Asc-task and Desc-task use significantly less computation time. This is because these ordering strategies do not require an additional process to retrieve information about the problem. Hence, considering both solution quality and computation time, it can be concluded that Asc-task and Desc-task (the second best known on average) should be selected for large instances because they produce solutions which are not much different from the other strategies but requiring significantly less computational time (48 % less on average).

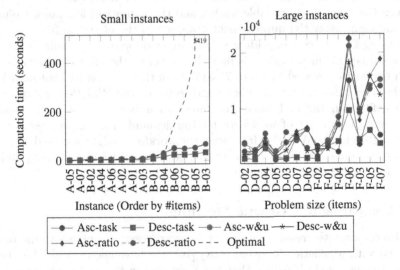

Fig. 5. Computation time (seconds) used in solving small and large instances. Each sub-figure corresponds to a problem size category (small and large). Instances are ordered by the problem size (#items) which is the summation of #workers and #tasks. Each graph presents the computation time used by the decomposition method with the different ordering strategies (line with markers) and the time used for producing the overall optimal solution (dashed line) when possible.

4.4 Geographical Decomposition with Neighbour Workforce

The study on the proposed decomposition method shows some limitation on tackling the working regions constraint. Recall that in our problem we consider the regions constraint to be soft. Assigning a worker to a region other than his/her own is allowed by penalised. However, the geographical decomposition technique described above enforces the regions constraint because each sub-problems corresponds to one region. When solving each sub-problem, the solver can only use the workers included in the sub-problem. Adding neighbour workers from nearby regions increases the feasible region by treating the regions constraint as soft. We only add workers to those regions in which the number of workers is less than the number of locations. Also, in order to maintain the size of the sub-problem manageable we add only enough workers to fill that difference. These additional workers are selected based on the distance from their departure location to the region, hence considered neighbour workforce. But also, the additional workers are selected from those with the highest set of skills and qualifications to be eligible to work in most of the tasks. This process of adding neighbour workforce is done before the process of further splitting a sub-problem. The only instances that require this process of adding neighbour workforce are instances WSRP-D and WSRP-F as presented in Table 5. For each instance, the table shows in columns two and six, the number of regions that required additional workers. Columns three and seven give the average ratio between the number of available worker and the number of locations. Columns four and eight show the improvement obtained in the objective function value when using this process of adding neighbour workforce. The result shows that additional neighbour workforce is more beneficial to the WSRP-F instances for which the cost decreased by up to 75.63 % from the solution without additional neighbour workforce. On the other hand, some of the WSRP-D instances did not benefit from the additional workforce, an indication that such instances have the right amount of workforce for the demand. This experimental result suggests that in the WSRP-F instances, the workforce might not be distributed well across regions according to the demanded tasks, which then causes problems when decomposing the problem by regions.

4.5 Comparing to a Genetic Algorithm

We also compare the results obtained by our proposed techniques to the results obtained with a genetic algorithm (GA) [16]. We have chosen a GA because it is a well-known meta-heuristic that has been proven to provide good solutions for both scheduling [18,25] and routing [19,28] problems.

We developed a straightforward GA implementation with uniform mutation crossover using a mutation rate $\alpha = \frac{1}{|T|}$ [16], binary tournament selection, population of 100 individuals where the 10 % best individuals are kept on the offspring population and time limit as stopping condition. Additionally, to avoid getting stuck in local optima and early convergence, we introduced a reset mechanism, that after 10 generations without improvement, the bottom half (the less fit

Table 5. Instances with understaffed regions. The second and sixth columns show the number of understaffed regions. The third and seventh columns show the average workforce/locations ratio. The fourth and eight columns show average decrease in the objective function value (improvement) obtained from having the additional neighbour workforce.

Instance	#Regions	Ratio	Decrease	Instance	#Regions	Ratio	Decrease
D-01	5	73.76 %	41.17 %	F-01	13	39.51 %	70.57 %
D-02	4	75.04 %	50.78 %	F-02	18	40.92 %	66.79 %
D-03	5	67.13 %	−5.87 %	F-03	22	31.87 %	67.33 %
D-04	5	76.77 %	0 %	F-04	17	40.09 %	70.04 %
D-05	4	73.08 %	0 %	F-05	19	37.97 %	56.29 %
D-06	5	64.72 %	0.10 %	F-06	13	44.58 %	75.63 %
D-07	7	69.31 %	7.34 %	F-07	23	33.73 %	53.57 %

#Regions is number of understaffed regions.
Ratio is average of proportion between workers and locations.
Decrease is the average improvement on the objective function value and is calculated as $\frac{(originalObj - addedWorkerObj)}{originalObj}$.

individuals) of the population is randomly re-generated acting as a population diversity procedure.

Problem-specific knowledge was introduced only at the chromosome encoding. In order to eliminate time conflicts and restrict the exploration to the feasible region of the solution space, we employed an indirect encoding. This encoding is composed of an array of integers such that the indexes correspond to the tasks (note that a task requiring multiple workers would have multiple indexes). The contents of each element in the array represents the k_{th} worker that is skilled to perform the task but that also represents a feasible assignment with no time conflict. If there is no such worker, than the task is left unassigned.

We now compare the results obtained with the proposed decomposition methods and the GA. Because of the stochastic nature of the GA, we ran the GA eight times on each instance and calculated the average solution quality obtained from all these runs. For the instance sets WSRP-A and WSRP-B, we employed the strategies Asc-task and Desc-task without additional neighbour workforce and we set a 15 min time limit for the GA. For the larger instance sets WSRP-D and WSRP-F, the additional neighbour workforce process was applied to the decomposition methods and the GA was ran for two hours. Table 6 presents the summary of the experimental results. Overall, we can see the GA shows better performance. This is clearer on the larger instances where the GA gives all best solutions on WSRP-F. However, it is important to mention that on such scenarios the decomposition technique took less computational time than the time limit given to the GA, except for instances WSRP-F-05 and WSRP-F-06 (see Fig. 5).

Table 6. Comparison of results obtained by the decomposition algorithms and the genetic algorithm on 28 instances.

	Asc-task	Desc-task	GA		N-A-task	N-D-task	GA
A-01	4.33	5.65	**3.92**	D-01	1168.56	**240.44**	337.09
A-02	3.53	4.53	**3.44**	D-02	259.99	**254.44**	324.82
A-03	7.54	10.65	**4.55**	D-03	2,933.84	3,212.32	**434.00**
A-04	**1.54**	3.09	1.75	D-04	**312.44**	418.89	386.26
A-05	2.77	3.54	**2.47**	D-05	408.42	**243.89**	356.47
A-06	**3.55**	3.74	3.70	D-06	307.05	1,410.77	**435.03**
A-07	5.25	4.81	**3.84**	D-07	949.55	753.28	**421.93**
B-01	2.00	**1.79**	1.82	F-01	18,046	22,020	**2,592.56**
B-02	1.94	1.89	**1.79**	F-02	25,712	25,654	**2,652.67**
B-03	2.25	**2.06**	2.27	F-03	50,944	33,686	**1,372.24**
B-04	2.29	**2.21**	2.49	F-04	29,965	34,070	**2,064.93**
B-05	5.60	4.74	**3.38**	F-05	59,363	41,433	**1,092.35**
B-06	2.62	2.52	**2.16**	F-06	20,853	22,038	**1,439.86**
B-07	4.30	4.06	**2.37**	F-07	109,398	66,969	**4,419.73**

Asc-task and **Desc-task** - GDCA ordered by ascending and descending task.
N-A-task and **N-D-task** - decomposition with neighbour workforce ordered by ascending and descending task.
GA - Genetic Algorithm.

The results obtained for the WSRP-F instances indicate that there is a drawback when the problem is decomposed into too small sub-problems. Our decomposition methods control the maximum size of each sub-problem to be the same for every instance, hence more tasks were left unassigned. As we mentioned before, the larger regions were split further into several small sub-problems which share the same workforce. The decomposition can be improved by generating a better cluster based on location and also uniformly distributed. Another challenge is choosing and clustering workforce which would improve the computation efficiency and workforce utilisation.

5   Conclusion and Future Work

A tailored mixed integer programming model for real-world instances of a workforce scheduling and routing problem is presented. The model is constructed by incorporating various constraints from the literature while also adding working region constraints to the formulation. It is usually the case that models in the literature for this type of problem are presented but their solution is provided using alternative methods such as heuristics because solving the model using mathematical exact solvers is computationally challenging. An approach using geographical decomposition with conflict avoidance is proposed here to tackle workforce scheduling and routing problems while still harnessing the power of exact solvers. The proposed decomposition method allows us to tackle real-world sized problems for which finding the overall optimal solution requires extensive computation time. However, the solution quality fluctuates when changing the order to tackle the sub-problems defined by the geographical regions. Exploring all permutation orders to find the one producing the best results is not practical for larger problems (e.g. more than 6 geographical regions). In this work, six ordering strategies are proposed for obtaining high-quality solutions within acceptable computation time. We also implemented a process of adding neighbour workforce to understaffed regions which helped to obtain better results. We also implemented a standard genetic algorithm to compare the results produced by the proposed decomposition methods. The experimental results indicate that although the decomposition methods produce some best results, it is outperformed by the GA on several instances. However, the decomposition methods are still faster in terms f computation time. Our future research will explore ways to improve the decomposition methods by re-defining sub-problems through automated clustering in order to find well separated regions that improves the solution procedure.

Acknowledgements. Special thanks to the Development and Promotion for Science and Technology talents project (DPST, Thailand) who providing partial financial support.

References

1. Bredström, D., Rönnqvist, M.: Combined vehicle routing and scheduling with temporal precedence and synchronization constraints. Eur. J. Oper. Res. **191**(1), 19–31 (2008)
2. Akjiratikarl, C., Yenradee, P., Drake, P.R.: PSO-based algorithm for home care worker scheduling in the UK. Comput. Ind. Eng. **53**(4), 559–583 (2007)
3. Angelis, V.D.: Planning home assistance for AIDS patients in the City of Rome, Italy. Interfaces **28**, 75–83 (1998)
4. Barrera, D., Nubia, V., Ciro-Alberto, A.: A network-based approach to the multi-activity combined timetabling and crew scheduling problem: workforce scheduling for public health policy implementation. Comput. Ind. Eng. **63**(4), 802–812 (2012)
5. Benders, J.: Partitioning procedures for solving mixed-variables programming problems. Numer. Math. **4**(1), 238–252 (1962)
6. Bertels, S., Torsten, F.: A hybrid setup for a hybrid scenario: combining heuristics for the home health care problem. Comput. Oper. Res. **33**(10), 2866–2890 (2006)
7. Borsani, V., Andrea, M., Giacomo, B., Francesco, S.: A home care scheduling model for human resources. In: 2006 International Conference on Service Systems and Service Management pp. 449–454 (2006)
8. Bredstrom, D., Ronnqvist, M.: A branch and price algorithm for the combined vehicle routing and scheduling problem with synchronization constraints. NHH Department of Finance & Management Science Discussion Paper No. 2007/7, February 2007

9. Castillo-Salazar, J., Landa-Silva, D., Qu, R.: Workforce scheduling and routing problems: literature survey and computational study. Ann. Oper. Res. **78**, 1–29 (2014)
10. Castro-Gutierrez, J., Landa-Silva, D., Moreno, P.J.: Nature of real-world multi-objective vehicle routing with evolutionary algorithms. In: 2011 IEEE International Conference onSystems, Man, and Cybernetics (SMC), pp. 257–264 (2011)
11. Cordeau, J.F., Stojkovic, G., Soumis, F., Desrosiers, J.: Benders decomposition for simultaneous aircraft routing and crew scheduling. Transp. Sci. **35**(4), 375–388 (2001)
12. Dantzig, G.B., Ramser, J.H.: The truck dispatching problem. Manage. Sci. (pre-1986) **6**(1), 80–91 (1959)
13. Dohn, A., Esben, K., Jens, C.: The manpower allocation problem with time windows and job-teaming constraints: a branch-and-price approach. Comput. Oper. Res. **36**(4), 1145–1157 (2009)
14. Eveborn, P., Ronnqvist, M., Einarsdottir, H., Eklund, M., Liden, K., Almroth, M.: Operations research improves quality and efficiency in home care. Interfaces **39**(1), 18–34 (2009)
15. Feillet, D.: A tutorial on column generation and branch-and-price for vehicle routing problems. 4OR **8**(4), 407–424 (2010). http://dx.doi.org/10.1007/s10288-010-0130-z
16. Goldberg, D.E.: Genetic Algorithms. Pearson Education (2006). ISBN: 9788177588293
17. Hart, E., Sim, K., Urquhart, N.: A real-world employee scheduling and routing application. In: Proceedings of the 2014 Conference Companion on Genetic and Evolutionary Computation Companion, GECCO Comp 2014, pp. 1239–1242. ACM, New York (2014)
18. Husbands, P.: Genetic algorithms for scheduling. Intell. Simul. Behav. (AISB) Q. **89**, 38–45 (1994)
19. Jeon, G., Leep, H.R., Shim, J.Y.: A vehicle routing problem solved by using a hybrid genetic algorithm. Comput. Ind. Eng. **53**(4), 680–692 (2007)
20. Kergosien, Y., Lente, C., Billaut, J.C.: Home health care problem, an extended multiple travelling salesman problem. In: Proceedings of the 4th Multidisciplinary International Scheduling Conference: Theory and Applications (MISTA 2009), Dublin, Ireland, pp. 85–92 (2009)
21. Landa-Silva, D., Wang, Y., Donovan, P., Kendall, G., Way, S.: Hybrid heuristic for multi-carrier transportation plans. In: The 9th Metaheuristics International Conference (MIC 2011), pp. 221–229 (2011)
22. Liu, R., Xie, X., Garaix, T.: Hybridization of tabu search with feasible and infeasible local searches for periodic home health care logistics. Omega **47**, 17–32 (2014)
23. Mankowska, D., Meisel, F., Bierwirth, C.: The home health care routing and scheduling problem with interdependent services. Health Care Manage. Sci. **17**(1), 15–30 (2014)
24. Mercier, A., Cordeau, J.F., Soumis, F.: A computational study of Benders decomposition for the integrated aircraft routing and crew scheduling problem. Comput. Oper. Res. **32**(6), 1451–1476 (2005)
25. Mesghouni, K., Hammadi, S.: Evolutionary algorithms for job shop scheduling. Int. J. Appl. Math. Comput. Sci. **2004**, 91–103 (2004)
26. Perl, J., Daskin, M.S.: A warehouse location-routing problem. Transp. Res. Part B Methodol. **19**(5), 381–396 (1985)

27. Pillac, V., Gueret, C., Medaglia, A.: On the dynamic technician routing and scheduling problem. In: Proceedings of the 5th International Workshop on Freight Transportation and Logistics (ODYSSEUS 2012), Mikonos, Greece, p. 194, May 2012
28. Potvin, J.Y.: Evolutionary algorithms for vehicle routing. Technical report 48, CIRRELT (2007)
29. Ralphs, T.K., Galati, M.V.: Decomposition Methods for Integer Programming. Wiley Encyclopedia of Operations Research and Management Science. Wiley, New York (2010)
30. Rasmussen, M.S., Justesen, T., Dohn, A., Larsen, J.: The home care crew scheduling problem: preference-based visit clustering and temporal dependencies. Eur. J. Oper. Res. **219**(3), 598–610 (2012)
31. Reimann, M., Doerner, K., Hartl, R.F.: D-Ants: savings based ants divide and conquer the vehicle routing problem. Comput. Oper. Res. **31**(4), 563–591 (2004)
32. Trautsamwieser, A., Hirsch, P.: Optimization of daily scheduling for home health care services. J. Appl. Oper. Res. **3**, 124–136 (2011)
33. Vanderbeck, F.: On Dantzig-Wolfe decomposition in integer programming and ways to perform branching in a branch-and-price algorithm. Oper. Res. **48**(1), 111 (2000)
34. Vanderbeck, F., Wolsey, L.: Reformulation and decomposition of integer programs. In: Junger, M., et al. (eds.) 50 Years of Integer Programming 1958–2008, pp. 431–502. Springer, Heidelberg (2010)

Local Search Based Metaheuristics for Two-Echelon Distribution Network with Perishable Products

Sona Kande[1,2(✉)], Christian Prins[1], Lucile Belgacem[2], and Benjamin Redon[2]

[1] ICD-LOSI, University of Technology of Troyes,
12 rue Marie Curie, CS 42060, 10004 Troyes Cedex, France
sona.kande,christian.prins@utt.fr
[2] FuturMaster, 1 cours de l'ile seguin, 92100 Boulogne-Billancourt, France
{sona.kande,lucile.belgacem,benjamin.redon}@futurmaster.com

Abstract. This article presents a planning problem in a distribution network incorporating two levels inventory management of perishable products, lot-sizing, multi-sourcing and transport capacity with a homogeneous fleet of vehicles. A mixed integer linear programming (MILP) and a greedy heuristic are developed to solve this real planning problem. There are some instances for which the solver cannot give a good lower bound within the limited time and for other instances it takes a lot of time to solve MILP. The greedy heuristic is an alternative to the mixed integer linear program to quickly solve some large instances taking into account original and difficult constraints. For some instances the gap between the solutions of the solver (MILP) and the heuristic becomes quite significant. The variable neighborhood descent (VND), the iterated local search (ILS) and the multi-start iterated local search (MS-ILS) are implemented. These methods are included in an APS (Advanced Planning System) and compared with a MILP solver. The instances are derived from actual data or built using a random generator of instances to have wider diversity for computational evaluation. The VND significantly improves the quality of solutions.

Keywords: Distribution network · Multi-start iterated local search · Local search · Lot-sizing · Metaheuristic · Multi-echelon inventory · Multi-sourcing · Perishable product · Supply planning · Transport capacity · Variable neighborhood descent

1 Introduction

In the problem under study, warehouses provide finished perishable products to various distribution centers with a homogeneous fleet of vehicles. A distribution center may be supplied by several warehouses. The choice of sourcing (warehouse) is determined by the availability of products in warehouses inventory, the fleet of vehicles, and the transport costs on routes. The goal is to come up with a compromise between the transportation costs, the warehouses and distribution

© Springer International Publishing Switzerland 2015
D. de Werra et al. (Eds.): ICORES 2015, CCIS 577, pp. 212–231, 2015.
DOI: 10.1007/978-3-319-27680-9_13

centers inventory costs, the loss due to products that are outdated. Furthermore, we comply with: flow conservation, inventory conservation at warehouses and distribution centers, capacity constraints (limited fleet of vehicles on each route) and supply constraints (lot-sizing, minimum order quantities and dates).

The remainder of this article is structured as follows. In Sect. 2, we present a literature review. The problem addressed in the paper is described in Sect. 3 and formulated as a mixed integer linear programming in Sect. 4. We explain how the metaheuristics are implemented in Sect. 5. The Sect. 6 is dedicated to computational evaluation. The last Sect. 7 gives the main contribution of this paper and the next steps of the work.

2 Literature Review

Hansen and Mladenović [5] propose the variable neighborhood search (VNS) in which several neighborhoods are successively used. VNS does not follow a single trajectory but explores increasingly distant neighbors of the incumbent. It jumps from this solution to a new one in case of improvement. Local search is used to get from these neighbors to local optima. VNS is based on the principle of systematically exploring several different neighborhoods, combined with a perturbation move (called shaking in the VNS literature) to escape from local optima. Variable neighborhood descent (VND) is essentially a simple variant of VNS, in which the shaking phase is omitted. Therefore, contrary to VNS, VND is usually deterministic.

Iterated local search (ILS) method is a popular heuristic in the literature. Lourenço et al. [7] present a tutorial about the ILS method and its application to some problems such as the travelling salesman problem and the scheduling, and compare the ILS with other metaheuristics; also with another recent paper [8]. Cuervo et al. [1] recently propose an ILS for the vehicle routing problem with backhauls (VRPB). It is an extension of the VRP that deals with two types of customers: the consumers (linehaul) that request goods from the depot and the suppliers (backhaul) that send goods to the depot. New best solutions have been found by the authors for two instances in one of the benchmark sets.

The greedy randomized adaptive search procedure (GRASP) [3,4] is a memory-less multi-start method in which local search is applied to different initial solutions constructed with a greedy randomized heuristic. Villegas et al. [13] propose two metaheuristics based on greedy randomized adaptive search procedures (GRASP), variable neighborhood descent (VND) and evolutionary local search (ELS). The problem under study is the single truck and trailer routing problem with satellite depots (STTRPSD): a vehicle composed of a truck with a detachable trailer serves the demand of a set of customers reachable only by the truck without the trailer. The accessibility constraint implies the selection of locations to park the trailer before performing the trips to the customers. The computational experiment shows that a multi-start evolutionary local search (GRASP-ELS) outperforms a GRASP/VND. Moreover, it obtains competitive results when applied to the multi-depot vehicle routing problem, that can be seen as a special case of the STTRPSD. The GRASP-ELS is also used by Duhamel et al. [2] for an extension of the capacitated vehicle routing problem where customer demand is composed of two-dimensional weighted items (2L-CVRP). The results show that their method is highly efficient and outperforms the best previous published methods on the topic.

A multi-start iterated local search (MS-ILS) is developed by Nguyen et al. [10] for the two-echelon location-routing problem (LRP-2E). They use three greedy randomized heuristics (cyclically to get initial solutions), the variable neighborhood descent and a tabu list. MS-ILS is also used by Michallet et al. [9] for the periodic vehicle routing problems with time spread constraints on services. Their paper addresses a real-world problem. The time windows must be absolutely respected and the hours of any two visits to the same customer must differ by a given time constant. They make evaluations on instances derived from classical benchmarks for the vehicle routing problem with time windows, and on two practical instances. For a particular case with a single period (the vehicle routing problem with soft time windows), MS-ILS competes with two published algorithms and improves six best known solutions.

We have not found any paper that proposes these methods for the problem under study.

3   Problem Description

The distribution network considered includes two levels: warehouses that provide perishable products to several distribution centers (see Fig. 1). A distribution center may be supplied by several warehouses. The choice of the sourcing (warehouse) is determined by the products availability in warehouses inventory, transport capacity and transport costs on the routes. The out of stock is permitted at the distribution centers; unmet demand may be postponed but is penalized by a cost. Stock levels are taken into account in the end of period. Every product shipped in a route has an associated cost of transport which determines the priority level of the warehouse. For each product-site (product-warehouse and product-distribution center) a stock policy is defined by three elements:

- a variable target stock in time: ideally reach target;
- a variable maximum stock in time: limit beyond which there is an overstock and product poses a risk of obsolescence;
- a variable minimum stock in time: reserve to cope with fluctuations in demand.

Products are made in batches to the warehouse. To differentiate with the concept of design in batches, "batch" stands for the set of product characterized by an amount of one type of product, the input date into the stock and expiry date. For each warehouse the batches of products are ordered in ascending order of their expiry dates and consumption occurs in this order. In addition, a freshness contract of product is made between the distribution center and their customers. It is a remaining life at least at the delivery time to the customer (at the period of distribution center requirement). A distribution center requirement

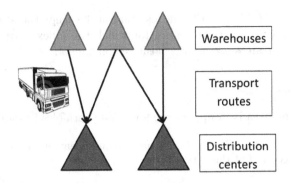

Fig. 1. Two echelon distribution network with multi-sourcing.

can be satisfied by several batches of the same product. When a batch is not consumed before its expiry date then there is a loss of the remaining amount. It is penalized by a cost called waste cost.

Constraints of lot size and a minimum supply quantity must be respected. The minimum quantity expresses an economic quantity necessary for profitable order launch and reception operations. The lot size constraint corresponds sending by pallets for the transport. The combination of these two constraints (lot size and minimum quantity) makes this much more complex than classical multi-echelon inventory management problems. A shipping calendar (opening or closing depending on the period) is also defined on each transport route.

The supply planning shall establish the quantities for each product and period to deliver to warehouses and distribution centers to satisfy at best the requirements (distribution centers) and reach the target stock (warehouses and distribution centers). The objective is to minimize the costs of transport, warehouses storage and distribution centers storage, the loss due to product obsolescence, out of stock penalties and tardiness penalties. We must respect the flow's coherence, the capacity constraints, the supply constraints (lot size, minimum quantity and dates) and seek to optimize the balancing of multi-echelon stock.

For reasons of flexibility, the user has the possibility of imposing quantity to supply inventory warehouses, distribution centers, also for transport. It is used to circumvent the various constraints in a critical situation. However, it can lead to inconsistencies in the calculation. During the execution of these heuristics, the inconsistencies are detected and reported without interrupting the supply planning calculation; which cannot be mixed with the MILP solver.

4   Mathematical Model

The problem can be formulated as a mixed integer linear programming. There are K warehouses (indexed by k), L distribution centers (indexed by l), P products (indexed by p), T number of periods of horizon of calculation (indexed by t), U number of periods of horizon of expiry dates (indexed by u) as

$u \in 1, ..., T + \max(DLU_{kp})$. All indexes start at 1 except the stock levels s_{lpt} (distribution center), s_{kpt} (warehouse) for which the index starts at 0 for the initial stock. Constraints use a large positive constant M.

4.1   Notations

The names of the data is in upper-case and those variables in lower-case.

$D_{lpt} \in IR^+$, $l \in L$, $p \in P$, $t \in T$: customer requirement of the product p at the distribution center l at the period t ;

Data for the Supply and the Shipment

$QT_{ipt}^{min} \in IN$, $i \in K \cup L$, $p \in P$, $t \in T$: minimum quantity of the product p to supply at the warehouse/distribution center i at the period t;

$CE_{kpt} \in \{0,1\}$, $k \in K$, $p \in P$, $t \in T$: shipment calendar; if $CE_{kpt} = 1$ the product p can be shipped from the warehouse k at the period t;

$CR_{lpt} \in \{0,1\}$, $l \in L$, $p \in P$, $t \in T$: reception calendar; if $CR_{lpt} = 1$ the product p can be receipt at the distribution center l at the period t;

$DLU_{kp} \in IN$: number of periods the quantity of the product p, supplied at the warehouse k, can be used : after this lead time the product is expired ;

$MCL_{lpt} \in IN$: minimum customer life is a remaining lifetime, at least, when the product p is delivered to the customer from the distribution center l at the period t);

$FPO_{ipt} \in IR^+$: firm planned order is a quantity of the product p imposed by the user at the warehouse/distribution center i at the period t;

$IsFPO_{ipt} \in \{0,1\}$: $IsFPO_{ipt} = 1$ if there is a quantity of the product p imposed by the user at the warehouse/distribution center i at the period t, so we can not supply a different quantity for the same product at this time (the value zero is considered);

$QEID_{lpt} \in IR^+$: quantity of the product p receipt at the distribution center l at the period t from a external sourcing.

Data for the Inventory

$S_{ip}^{init} \in IR^+$, $i \in K \cup L$, $p \in P$: initial stock level of the product p at the warehouse/distribution center i ;

$S_{ipt}^{obj} \in IR^+$, $i \in K \cup L$, $p \in P$, $t \in T$: target stock level of the product p at the warehouse/distribution center i at the period t;

$S_{ipt}^{max} \in IR^+$, $i \in K \cup L$, $p \in P$, $t \in T$: maximum stock level of the product p at the warehouse/distribution center i at the period t ;

$S_{ipt}^{min} \in IR^+$, $i \in K \cup L$, $p \in P$, $t \in T$: minimum stock level of the product p at the warehouse/distribution center i at the period t ;

$P_{ipt}^{>max} \in IR^+$, $i \in K \cup L$, $p \in P$, $t \in T$: unit penalty of overstock if we exceed the maximum stock level of the product p at the warehouse/distribution center i at the period t (per unit of surplus stock);

$P_{ipt}^{>obj} \in IR^+$, $i \in K \cup L$, $p \in P$, $t \in T$: unit earliness penalty when the stock level, of the product p at the warehouse/distribution center i at the period t, is between the target stock and the maximum stock;

$P_{ipt}^{<obj} \in IR^+$, $i \in K \cup L$, $p \in P$, $t \in T$: unit tardiness penalty when the stock level, of the product p at the warehouse/distribution center i at the period t, is between the minimum stock and the target stock (by missing unit to reach the target stock);

$P_{ipt}^{<min} \in IR^+$, $i \in K \cup L$, $p \in P$, $t \in T$: unit penalty applied when the stock level, of the product p at the warehouse/distribution center i at the period t, is between 0 and the minimum stock (by missing unit to reach the minimum stock);

$P_{lpt}^{rupt} \in IR^+$, $l \in L$, $p \in P$, $t \in T$: unit penalty of out of stock applied when the stock level, of the product p at the distribution center l at the period t, is under 0 (the out of stock is only permitted at the distribution centers);

$P_{kpt}^{obs} \in IR^+$, $k \in K$, $p \in P$, $t \in T$: unit penalty of obsolescence of the product p at the warehouse k at the period t (the lifetime is is exceeded).

Data for the Transport

$QTA_{klpt}^{min} \in IN$: minimum quantity of the product p shipped on edge (k, l) at the period t;

$TL_{klp} \in IN$: lot size for the shipment of the product p on the edge (k, l) in units of product;

$CT_{klpt} \in \{0, 1\}$: $CT_{klpt} = 1$ if the transport of the product p is permitted on the edge (k, l) at the period t;

$CA_{klt} \in IN$, $k \in K$, $l \in L$, $t \in T$: available capacity on the edge (k, l) at the period t, expressed as the number of pallets;

$CoPa_p \in IR^+$, $p \in P$: conversion factor quantity $\rightarrow$ pallets of the product p;

$HV_{kl} \in IR^+$, $k \in K$, $l \in L$: cost of operating a vehicle on the edge (k, l) (counted only once even if the vehicle travelled several days);

$HT_{klpt} \in IR^+$, $k \in K$, $l \in L$, $p \in P$, $t \in T$: unit cost of the shipment of the product p on the edge (k, l) at the period t;

$LT_{klt} \in IN$, $k \in K$, $l \in L$, $t \in T$: lead time of transport on the edge (k, l) at the period t;

$NP^{max} \in IN$: maximum number of pallets a vehicle can be loaded (same for all vehicles);

$PV^{max} \in IR^+$: maximum loading weight of a vehicle (same for all vehicles);

$PU_p \in IR^+$, $p \in P$: unit weight of the product p;

$FPOA_{klpt} \in IR^+$: quantity imposed to ship the product p on the edge (k, l) at the period t (useful for imposing an origin and a destination);

$IsFPOA_{klpt} \in \{0, 1\}$: $IsFPOA_{klpt} = 1$ if there is a quantity imposed to ship the product p on the edge (k, l) at the period t ($[FPOA_{klpt}$;

v_{klt}^{max}, $k \in K$, $l \in L$, $t \in T$: number of vehicles used on the edge (k, l) at the period t.

Variables for the Supply and the Shipment

qi_{ipt}, $i \in K \cup L$, $p \in P$, $t \in T$: quantity of the product p supplied at warehouse/distribution center i at the period t;

$isqi_{ipt} \in \{0, 1\}$, $i \in K \cup L$, $p \in P$, $t \in T$: $isqiw_{ipt} = 1$ if a non-zero quantity of the product p is supplied at warehouse/distribution center i at the period t;

qo_{kpt}, $k \in K$, $p \in P$, $t \in T$: quantity of the product p shipped by warehouse k at the period t (output quantity);

qp_{kpt}, $k \in K$, $p \in P$, $t \in T$: quantity of the product p lost due to obsolescence at the warehouse k at the period t;

$n_{klpt} \in IN$, $k \in K$, $l \in L$, $p \in P$, $t \in T$: number of lots that form the quantity shipped on the edge (k, l) at the period t;

qe_{klpt}, $k \in K$, $l \in L$, $p \in P$, $t \in T$: quantity of the product p shipped on the edge (k, l) at the period t;

$isqe_{klpt} \in \{0, 1\}$, $k \in K$, $l \in L$, $p \in P$, $t \in T$: $isqe_{klpt} = 1$ if a non-zero quantity of the product p is shipped on the edge (k, l) at the period t;

qed_{klpu}, $k \in K$, $l \in L$, $p \in P$, $u \in U$: total quantity of the product p shipped on the edge (k, l) and whose expiration date is u;

$qedp_{klptu}$, $k \in K$, $l \in L$, $p \in P$, $t \in T$, $u \in U$: quantity of the product p shipped on the edge (k, l) at the period t and whose expiration date is u;

qoi_{klpt}, $k \in K$, $l \in L$, $p \in P$, $t \in T$: quantity of the product p shipped by the warehouse k and receipt at the distribution center l at the period t .

Variables for the Inventory

sed_{kpu}: quantity of the product p and whose expiration date is u at the warehouse k;

s_{ipt}, $i \in K \cup L$, $p \in P$, $t \in T$: stock level of the product p at the warehouse k or at the distribution center l at the period t;

s_{lpt}^{rupt}, $l \in L$, $p \in P$, $t \in T$: missing quantity to avoid the out of stock when the stock level is negative;

$s_{ipt}^{<min}$, $i \in K \cup L$, $p \in P$, $t \in T$: missing quantity to reach the minimum stock, from 0 until the minimum stock;

$s_{ipt}^{<obj}$, $i \in K \cup L$, $p \in P$, $t \in T$: missing quantity to reach the target stock, from the minimum stock until the target stock;

$s_{ipt}^{>obj}$, $i \in K \cup L$, $p \in P$, $t \in T$: quantity between the stock level and the target stock, from the target stock until the maximum stock;

$s_{ipt}^{>max}$, $i \in K \cup L$, $p \in P$, $t \in T$: quantity between the stock level and the maximum stock, when the stock level is over the maximum stock.

4.2   Stock Cost Formulation

Two cost functions, whose shape is shown in the Fig. 2, are defined to evaluate the inventory of the warehouses and the distribution centers. The function $f(s_{ipt})$ means the stock cost of the product p in the warehouse/distribution center i at time t. This is a piecewise linear convex function. It has a zero value when the stock level is equal to the target stock. This corresponds to a cost of storage and optionally of overstock (stock level greater than the maximum stock) when the stock level is above the target stock. When the stock level is lower than the target stock, it is the cost of delay and possible risk of out of stock (stock level below the minimum stock). For the distribution centers, a penalty for negative stock (cost of out of stock) is added.

Illustrated in the Fig. 3, the variables of gap are defined to linearise the piecewise linear functions. These are the quantities of gap for each piece. Cost functions of stock at the warehouses and the distribution centers may be formulated as follows:

$$f(s_{ipt}) = \left\{ \begin{array}{c} \left[P_{lpt}^{rupt} \cdot s_{lpt}^{rupt} \right] + P_{ipt}^{<min} \cdot s_{ipt}^{min} + P_{ipt}^{<obj} \cdot s_{ipt}^{<obj} \\ + P_{ipt}^{>obj} \cdot s_{ipt}^{>obj} + P_{ipt}^{>max} \cdot s_{ipt}^{>max} \end{array} \right.$$

Fig. 2. Stock policy and the penalties.

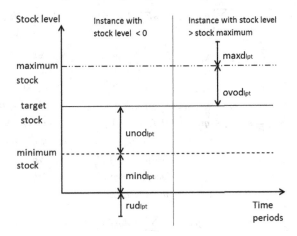

Fig. 3. Variables for linearisation of the piecewise linear functions.

4.3 MILP Formulation

Objective function:

$$\text{Min } Z = \qquad\qquad (1)$$

$$\underbrace{\sum_{k=1}^{K}\sum_{l=1}^{L}\sum_{t=1}^{T}\left[(HV_{kl}\cdot v_{klt}^{max}) + \sum_{p=1}^{P}(qe_{klpt}\cdot HT_{klpt})\right]}_{\text{transport cost}}$$

$$+ \underbrace{\sum_{i=1}^{K\cup L}\sum_{p=1}^{P}\sum_{t=1}^{T}f(s_{ipt})}_{\text{warehouses and distribution centers stock cost}}$$

$$+ \underbrace{\sum_{k=1}^{K}\sum_{p=1}^{P}\sum_{t=1}^{T}P_{kpt}^{obs}\cdot qp_{kpt}}_{\text{cost of loss due to obsolescence}}$$

subject to:

$$qe_{klpt} = FPOA_{klpt} \text{ if } IsFPOA_{klpt} = 1 \qquad \forall k \quad \forall l \quad \forall p \quad \forall t \qquad (2)$$

$$qe_{klpt} = n_{klpt} \cdot TL_{klp} \text{ if } IsFPOA_{klpt} = 0, TL_{klp} \neq 0 \qquad \forall k \quad \forall l \quad \forall p \quad \forall t \qquad (3)$$

$$qe_{klpt} \geq isqe_{klpt} \cdot (1 - IsFPOA_{klpt}) \cdot QTA_{klpt}^{min} \qquad \forall k \quad \forall l \quad \forall p \quad \forall t \qquad (4)$$

$$qe_{klpt} \leq isqe_{klpt} \cdot M \qquad \forall k \quad \forall l \quad \forall p \quad \forall t \qquad (5)$$

$$qe_{klpt} = qe_{klpt} \cdot CT_{klpt} \text{ if } IsFPOA_{klpt} = 0 \qquad \forall k \quad \forall l \quad \forall p \quad \forall t \qquad (6)$$

$$qoi_{klp,t+LT_{klt}} = qe_{klpt} \qquad \forall k \quad \forall l \quad \forall p \quad \forall t \qquad (7)$$

$$qi_{ipt} = FPO_{ipt} \text{ if } IsFPO_{ipt} = 1 \qquad \forall i \quad \forall p \quad \forall t \qquad (8)$$

$$qi_{ipt} \geq isqi_{ipt} \cdot (1 - IsFPO_{ipt}) \cdot QT_{ipt}^{min} \qquad \forall i \quad \forall p \quad \forall t \qquad (9)$$

$$qi_{ipt} \leq isqi_{ipt} \cdot M \qquad \forall i \quad \forall p \quad \forall t \qquad (10)$$

$$qi_{lpt} = CR_{lpt} \cdot \sum_{k=1}^{K} qoi_{klpt} \text{ if } IsFPO_{ipt} = 0 \qquad \forall l \quad \forall p \quad \forall t > LT_{klt} \qquad (11)$$

$$qo_{kpt} = CE_{kpt} \cdot \sum_{l=1}^{L} qe_{klpt} \qquad \forall k \quad \forall p \quad \forall t \qquad (12)$$

$$s_{ip0} = S_{ip}^{init} \qquad \forall i \quad \forall p \qquad (13)$$

$$s_{kpt} = s_{kpt-1} + qi_{kpt} - qo_{kpt} \qquad \forall k \quad \forall p \quad \forall t > 0 \qquad (14)$$

$$s_{lpt} = s_{lpt-1} + qi_{lpt} + QEID_{lpt} - D_{lpt} \qquad \forall l \quad \forall p \quad \forall t > 0 \qquad (15)$$

$$sed_{kpu} = 0 \qquad \forall k \quad \forall p \quad \forall u < DLU_{kp} \qquad (16)$$

$$sed_{kpu} = S_{kp}^{init} \qquad \forall k \quad \forall p \quad u = DLU_{kp} \qquad (17)$$

$$sed_{kpu} = qi_{kpt} \qquad \forall k \quad \forall p \quad \forall t$$

$$u = t + DLU_{kp} \qquad (18)$$

$$qe_{klpt} = \sum_{u=1}^{U} qedp_{klptu} \qquad \forall k \quad \forall l \quad \forall p$$

$$\forall t >= u - DLU_{kp} \qquad (19)$$

$$qed_{klpu} = \sum_{t=1}^{u} qedp_{klptu} \qquad \forall k \quad \forall l \quad \forall p \quad \forall u \qquad (20)$$

$$\sum_{l=1}^{L} qed_{klpu} \leq sed_{kpu} \qquad \forall k \quad \forall p \quad \forall u \qquad (21)$$

$$\sum_{u=1}^{U} qed_{klpu} \leq \sum_{t=1}^{T} D_{lpt} \qquad \forall k \quad \forall l \quad \forall p \quad \forall u \leq T$$

$$\forall t \leq u - MCL_{lpt} \qquad (22)$$

$$qp_{kpt} = sed_{kpu} - \sum_{l=1}^{L} qed_{klpu} \qquad \forall k \quad \forall p \quad \forall t = u \qquad (23)$$

$$CA_{klt} \geq v_{klt}^{max} \cdot NPmax \qquad \forall k \quad \forall l \quad \forall t \qquad (24)$$

$$v_{klt}^{max} \geq \frac{\sum_{p=1}^{P} qe_{klpt} \cdot COPA_p}{NPmax} \qquad \forall k \quad \forall l \quad \forall t \qquad (25)$$

$$v_{klt}^{max} \geq \frac{\sum_{p=1}^{P} (qe_{klpt} \cdot PU_p)}{PVmax} \qquad \forall k \quad \forall l \quad \forall t \qquad (26)$$

$$s_{lpt}^{rupt} \geq 0 - s_{lpt} \qquad \forall l \quad \forall p \quad \forall t \qquad (27)$$

$$s_{ipt}^{<min} [+s_{lpt}^{rupt}] \geq S_{ipt}^{min} - s_{ipt} \qquad \forall i \quad \forall p \quad \forall t \qquad (28)$$

$$s_{ipt}^{<obj} + s_{ipt}^{<min} [+s_{lpt}^{rupt}] \geq S_{ipt}^{obj} - s_{ipt} \qquad \forall i \quad \forall p \quad \forall t \qquad (29)$$

$$s_{ipt}^{>obj} + s_{ipt}^{>max} \geq s_{ipt} - S_{ipt}^{obj} \qquad \forall i \quad \forall p \quad \forall t \qquad (30)$$

$$s_{ipt}^{>max} \geq s_{ipt} - S_{ipt}^{max} \qquad \forall i \quad \forall p \quad \forall t \qquad (31)$$

$$s_{kpt}, sed_{kpu} \geq 0 \qquad \forall k \quad \forall p \quad \forall t \quad \forall u \qquad (32)$$

$$s_{lpt} \in IR \qquad \forall l \quad \forall p \quad \forall t \qquad (33)$$

$$n_{klpt} \in IN \qquad \forall k \quad \forall l \quad \forall p \quad \forall t \qquad (34)$$

$$qo_{kpt}, qp_{kpt}, qi_{ipt}, qe_{klpt}, qoi_{klpt}, qed_{klpu},$$
$$qedp_{klptu} \geq 0 \qquad \forall i \quad \forall k \quad \forall l \quad \forall p \quad \forall t \quad \forall u \qquad (35)$$

$$v_{klt}^{max} \in IN \qquad \forall k \quad \forall l \quad \forall t \qquad (36)$$

$$isqe_{klpt}, isqi_{ipt} \in \{0, 1\} \qquad \forall k \quad \forall l \quad \forall p \quad \forall t \qquad (37)$$

$$s_{lpt}^{rupt} \geq 0 \qquad \forall l \quad \forall p \quad \forall t \qquad (38)$$

$$s_{ipt}^{<min}, s_{ipt}^{<obj}, s_{ipt}^{>obj}, s_{ipt}^{>max} \geq 0 \qquad \forall i \quad \forall p \quad \forall t \qquad (39)$$

The objective function (1) includes the transport costs, the costs of stock for the warehouses and distribution centers, the cost of loss due to obsolescence. The constraints (2) and (8) ensure that the imposed quantities are respected. The lot size of the shipped quantities is respected via the constraint (3). Constraints (4), (5), (9) and (10) concern the minimum quantities for the receipt quantities at the distribution centers and the warehouses, and the shipped quantities on the routes. If there is not a imposed quantity, we can ship if the shipping calendar is open (6). The shipping quantity arrives at the distribution center after the lead time on the route (7). The receipt quantity at the distribution center corresponds to the sum of all the shipped quantities and the reception calendar must be opened (11). The output quantity at the warehouse is equal to the sum of all the shipped quantities and the shipping calendar must be opened (12). Flow conservation and the levels of inventory are expressed via constraints (13)–(15). The inventory stock by the expiration date is expressed via constraints (16)–(18). Constraints (19)–(21) concern the use of batches of quantities taking into account their expiry date. Constraints (22) guarantee that the freshness contract (minimum customer life) is respected. The lost quantities in the warehouse (the batches that are not consumed before expiry date) are calculated using constraint (23). Constraints (24)–(26) check that the transport capacity is not exceeded by the number of pallets and the total weight of shipped products. The linearisation of the variables of gap are made with constraints (27)–(31) for the stock at the distribution centers and the stock at the warehouses. The variables are defined in the lines (32)–(39).

There are some instances for which the solver cannot give a good lower bound within the limited time and for other instances it takes a lot of time to solve MILP. A greedy heuristic has been implemented as an alternative to the mixed integer linear program to quickly solve some large instances. For more precision about the greedy heuristic, see our previous article [6]. For some instances the gap between the solution provided by the solver (MILP) and the heuristic becomes quite significant. Metaheuristics are implemented to improve the quality of heuristic solutions.

5 Methods

5.1 Variable Neighborhood Descent

The different movements used in our VND procedure are more or less complex. Simple movements are used to increase or reduce the supplied amounts to the distribution centers and the warehouses. Complex movements can transfer amounts from a time period to another, from a distribution center to another and from a warehouse to another.

The decrease of the supplied quantities (movements 1 and 6) is designed to avoid overstock: when the stock level is above the target stock at product-site couples (product-warehouse and product-distribution center). When applied on a product-distribution center the quantity is spread across the flow; on the route and the sourcing (product-warehouse). The increase of supplied quantity (movements 3 and 7) reduce the gap when the stock level is below the target stock level at couples product-site. Amounts may be transferred between different periods of the horizon for all product-site couples (movements 2 and 9). The amount removed from the starting period is not necessarily equal to that added at the arrival time. An exchange can occur between two product-distribution center couples sharing the same transport route (movement 4). The interest of this movement is to find a balance between the stock levels of the different products in a distribution center using the same warehouse provider. Consumption of transport capacity is not optimized by the greedy heuristic. The quantity removed from the original product-distribution center may be different from the added amount. Different distribution centers supplying the same product from the same warehouse are also the subject of an exchange (movement 8). A particular movement (movement 5) is integrated to reduce the cost of loss due to product obsolescence in warehouses. When the removal is possible, the batch with outdated products is exchanged with one or several batches to meet the needs of product-distribution center couples. The constraints of date such as expiry date and contract freshness are always respected.

The VND stops if no movement improves solution cost or the time limit is reached or the percentage of improvement (IP) of solution cost between two successive iterations is less than 0.01 %. This gap is also a stop criterion for solver MILP resolution. The second criterion can reduce computational time without degrading solution quality. The formula of this gap is as follows:

$$IP = \frac{\text{last solution cost - solution cost}}{\text{solution cost}} \times 100 \qquad (40)$$

```
VND algorithm
    Cost(Sol) := heuristic solution cost;
    Initialise Number_Move;
    Repeat
      Cost(LastSol):= Cost(Sol);
      i:= 1;
      While (i <= Number_Move)
        For (Neighborhood(Sol,i))
            Sol':= LocalSearch N(Sol,i);
            If (Cost(Sol') < Cost(Sol))
               Sol:= Sol';
            EndIf;
        EndFor;
      EndWhile;
    Calculate IP;
Until(Cost(Sol)>=Cost(LastSol) or IP<=0.01 or Time limit reached).
```

5.2   Neighborhood Space Reduction

The neighborhood construction occurs on exploring three dimensions: the product-site couple, the time period and the quantity. For the transfer movements, exploration is done on 6 dimensions: two times the same triplet (product-site couple, time period and quantity) for the start and end situations. The neighborhood space is very large and its exploration can take a lot of time. The first time a movement is applied all neighborhood space is explored. An indicator is made on each modified product-site couple and its connections: others product-site couples with whom it has a link (for instance, the product-distribution center couples using the same route or the same sourcing). At the same time, the first time period of horizon, where there is a modification, is marked. The next time the movements are only applied on modified product-site couples and from the first modified time period of horizon, not through all the horizon of calculation. The exploration indicators in a neighborhood space are modified product-site couple and the first modified time period of the horizon.

Considering only the inventory cost, when the stock level is over the target stock there is no need to add quantities. The same reasoning can be done when the stock level is under the target stock: if the supplied quantity is decreased, inventory cost increases. These checks on the inventory state allow reducing neighborhood space for decrease and increase movements.

5.3   Iterated Local Search

For Lourenço et al. [7], the ILS has many of the desirable features of a metaheuristic: it is simple, easy to implement, robust, and highly effective. The essential idea of ILS lies in focusing the search not on the full space of solutions but on a smaller subspace defined by the solutions that are locally optimal for a given optimization engine. Each iteration, a copy of the best known solution is disturbed and the local search is generally used to improve it. The strength of the method is that the solutions generated are dependent and trace a path of local optima near each other in the space of solutions. ILS is also very fast: if the disturbance is slight enough, the new solution looks to the parent solution and the local search needs only few movements to re-optimize it.

```
ILS algorithm
    Execute VND (BestSol);
    Initialise shaking parameters k_min, k_max;
    Initialise maximum number of iterations max_iterator;
    k:= k_min;
    i:= 1;
    Repeat
        Sol:= BestSol;
        Execute movements of perturbation (Sol)
        execute VND (Sol)
        If (Cost(Sol) < Cost(BestSol))
            BestSol:= Sol;
            k := k_min;
        Else
            Sol:= BestSol;
            k := k + 1;
            If (k > k_max)
                    k := k_min;
            EndIf
        EndIf
    Until (i > max_iterator or Time limit reached).
```

In our method, the greedy heuristic is used as initial solution and variable neighborhood descent (VND) is used to improve this solution. We get a local optimum at the end of the VND. At each iteration, the movements of perturbation are executed to get a new solution from which the VND is applied. After the perturbation the new solution is different and less good than the local optimum but the VND can improve it and give a better solution. The ILS stops if a fixed number of iterations or the time limit are reached. The higher the number of iterations, the greater the computational time is.

Perturbation. Perturbation is a mechanism to escape from local optima. Our perturbation is powered by four movements. The first is the movement of removing obsolete batch for product-warehouse couples. Deleting obsolete batches can degrade the stock of the warehouse over several periods. This causes an increase in the stock cost and cancel the decrease in the waste cost. When this perturbation movement is applied the change is accepted even if the cost of the solution increases. The movement of adding quantity for product-warehouse couples can correct this effect in the next VND application. The second and third movements are movements of decreasing quantity, respectively for the product-distribution center couples and the product-warehouse couples. These movements are applied in VND and may cause deterioration of the stock over future periods. The fourth movement is used to exchange amounts between the product-distribution center couples sharing the same source. In some cases the available warehouse resources is restricted to satisfy all requirements of distribution centers. Some exchanges of amounts are made to modify the distribution made by the greedy heuristic and let VND improve the solution.

For each movement of perturbation, the list of product-site (product-distribution center couples or the product-warehouse couples) is randomly sorted. Each movement is applied on 10 % of the associated list so that the VND easily repairs the perturbation. The ILS is tested with a constant shaking rate of perturbation (10 %).

5.4   Multi-start Iterated Local Search

We were inspired by the papers of Prais and Ribeiro [11] and, Prins and Calvo [12] for the GRASP. Prais and Ribeiro propose reactive method GRASP for a problem in telecommunication. Prins and Wolfler Calvo also propose a GRASP which is repeatedly running a greedy randomized heuristic that improves with a local search. In our method, the MS-ILS method uses several runs of ILS method, which is developed with a reactive randomized heuristic(RRH) and the variable neighborhood descent (VND) method.

```
MS-ILS algorithm
   Initialise maximum number of runs: Number_run;
   Initialise maximum number of iterations: max_iterator;
   Execute 100 times the RRH to get statistics and identify the best settings;
   j:= 1;
      Repeat
         execute the ILS method with the RRH as initial solution;
      Until (j > Number_run or Time limit reached).
```

The heuristic is divided into three main procedures to better balance stock levels. The requirement of the distribution centers is first processed regardless of target stock: this is to determine the amount necessary to avoid out of stock (minimum supply process). Then, we seek to achieve the target stock. Finally, the surplus in the warehouses is deployed to the distribution centers. This division into three phases allows better management of cases of shortage or very limited capacity, since it seeks to ensure that all the distribution centers are supplied to meet requirement, before trying to establish the target stock.

Two types of randomization have been incorporated in the heuristic. This is the sort of list of couples product-warehouse and the product-distribution center.

Randomization of the List of Couples Product-Warehouse. In the process of minimum supply list couples product warehouse is sorted in "natural" order (each product-warehouse is indexed by its memory address). For each product-warehouse, the list of couples supplied product-distribution center is generated. The available quantity is shared in order to avoid out of stock (at couples product-distribution center). The purpose of this randomization is to test different courses from the list of couples product-warehouse. A parameter k is defined, it represents the number of elements of the set. A couple is randomly chosen among the k first items on the list of couples product-warehouse. The method is as follows:

- create a set with the first k elements;
- randomly select an item in the set;
- allocate quantities to avoid out of stock for couples product-distribution center which are mainly supplied from the selected product-warehouse;
- integrate the next element located at position k+1 in the list as a whole.

Randomization of the List of Couples Product-Distribution Center. In the procurement process to achieve the target stock of couples product-distribution center, the list of couples is sorted in order of the memory address. For each product-distribution center, the just in time supply is made from the main sourcing (product-warehouse): the level of priority is defined by unit cost of transportation and the lowest cost is associated to the transport route from the main warehouse. If the target stockt is not achieved at this step then other options as early and late supply and from other warehouses (secondary, tertiary, etc. ...) are evaluated and applied. We seek to disrupt the criticality associated with couples product-distribution center. The order of the list of couples product-distribution center is modified. This randomization is performed as follows:

- calculate the criticality of each product-distribution center: in the case of sorting criteria default value of 1 is assigned to each couple;
- randomly increase the value for each criticality of 0 to k %;
- sort the list of couples distribution center product in descending order of criticality values obtained.

6   Computational Evaluation

The metaheuristics (VND, ILS and MS-ILS) are implemented in C + + Visual Studio 2010 development environment. The MILP is solved with the solver CPLEX version 12.6. The instances (60) are tested on a server with an Intel Xeon 2.93 GHz processor and 48 GB of RAM.

6.1   Instances

The 20 first instances have been extracted from actual database representing real distribution networks for customers of FuturMaster, a french software publisher. The 40 other instances have been randomly generated to have wider diversity for computational evaluation. First database has provided 10 small instances with the number of products varies from 2 to 10, 2 warehouses, 2 distribution centers and horizon times is from 10 to 20 periods. The 10 instances from second database have the number of products that varies from 5 to 20, 11 warehouses, 145 distribution centers and horizon times varying from 10 to 20 periods. We have tried to evaluate greater instances with this database but when the number of products exceeds 20 the solver is out of memory. The random generator of instances has provided 40 instances with the number of products varying from 5 to 50, warehouses from 5 to 10, distribution centers from 10 to 20 and horizon times from 10 to 30 periods. In all instances, each product is perishable, with a lifetime, and a freshness contract (minimum customer life) is defined for each couple product-distribution center. To fix ideas, the MILP of the smaller instance contains 1,760 constraints and 2,569 variables and for the greater instance there are 921,712 constraints and 2,760,510 variables.

6.2   Results

The results of MILP solver and greedy heuristic (initial solution) for all instances are in Table 1. The default parameter of relative gap between lower bound (LB) and upper bound (UB), 0.01, is used as stop criterion. The computational time is in seconds. The gap between greedy heuristic solution cost (H) and lower bound of MILP (H/LB) and its computational times are given in Table 1:

$$\% \text{ Gap UB/LB} = \frac{\text{UB - LB}}{\text{LB}} \times 100 \tag{41}$$

$$\% \text{ Gap H/LB} = \frac{\text{H - LB}}{\text{LB}} \times 100 \tag{42}$$

We note that for the instances from the database 1 the MILP resolution finds a good solution with 0.001 as average gap between lower and upper bound for 29.51 s. For the instances extracted from the database 2 and from the random generator, the computational time of MILP resolution increases: in average 7427.13 s for database 2 and 1418.28 s for the instances generated randomly. For some instances of the random generator, the solver does not provide a good lower bound after 1 hour and half of calculation: 11 are not represented in the results and heuristic results are not compared with MILP solver. The greedy heuristic is very fast (in average less than 1 second) but for the instances of the database 2 (15.971 %) and the random generator (31.829 %) the average gap, between the LB and the cost of heuristic solution, becomes quite significant. The worst cases are due to the structure of the greedy heuristic.

Table 1. Results of MILP solver and greedy heuristic.

Instances from	MILP solver		Heuristic (H)	
	%Gap UB/LB	Time (s)	%Gap H/LB	Time (s)
Database 1	0.001	29.51	0.010	0.02
Database 2	0.072	7427.13	15.971	0.66
Random generator	0.519	1418.28	31.829	0.04
Average for all	0.322	2361.15	22.233	0.16

The results of the methods VND, ILS and MS-ILS are in Table 2. The gaps between the costs of solutions of VND (ILS and MS-ILS) and the lower bound of MILP, are calculated as follows:

$$\% \text{ Gap VND/LB} = \frac{\text{VND - LB}}{\text{LB}} \times 100 \tag{43}$$

$$\% \text{ Gap ILS/LB} = \frac{\text{ILS - LB}}{\text{LB}} \times 100 \tag{44}$$

$$\% \text{ Gap MS-ILS/LB} = \frac{\text{MS-ILS - LB}}{\text{LB}} \times 100 \tag{45}$$

VND Results. The instances from the database 1 are slightly improved by the VND because the gap H/LB was very small. The initial solution found by the greedy heuristic gives very good solutions: the average gap between the heuristic solution and the MILP solver solution is 0.010 (see Table 1). For the instances from the database 2, the costs of the solutions decrease by 14 % and the gap between the VND solution cost and the lower bound of MILP resolution is 1.184 %. The movements significantly improve these instances because the weaknesses of greedy heuristic are corrected. The average computational time of the VND is 142.10 s; it is much shorter than MILP resolution which is 7427.13 s. The solution costs of the instances from the random generator decrease, even more, by 25 % compared with the solution of the greedy heuristic. The average gap, between the LB and the VND solution costs, is 6.743 % for a average computational time of 292.52 s.

For all evaluated instances, the average gap between the lower bound of MILP resolution and the VND is 4.234 %. The gap decreases by 17.9 % when the VND is applied (see Fig. 4) and the average computational time is much shorter than MILP resolution. The VND (202.70 s in average) is 11 times faster than the solver (2361.15 s in average).

ILS Results. The ILS gives the same results as the VND for the instances from the database 1; the solutions are very good. The instances from the database 2

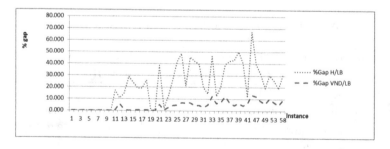

Fig. 4. Comparing % Gap VND/LB and % Gap H/LB.

are slightly improved by ILS: the quality of the solutions is improved by 0.059 %, and the average computational time increase with the ILS method (1324.90). The VND found a good solution yet; these instances don't need the ILS. The instances randomly generated present more difficulties as situations of shortage. For some instances, the available quantities in a warehouse are fixed over a number of periods: the available resource is randomly generated by taking as upper bound X% of the target stock. The costs of the ILS solutions decrease, in average, by 1.107 % compared with the solution of the VND. The average computational time of ILS is 964.97 s for the instances from the random generator. It is less shorter than MILP resolution which is 1418.28 s. The next step is to reduce the computational time of ILS by changing the rate of perturbation and to test more iterations.

For all the evaluated instances, the average gap between the lower bound of MILP resolution and the ILS is 3.567 %. The gap decreases by 0.667 % compared with the solution of VND. The computational time of ILS (851.23 s) is much shorter than MILP resolution (2361.15 s). The costs of the solutions are quite significant, the gain provided by the ILS can be significant in actual costs. For one of instances, the cost of the solution decrease by 0.056 % if we compare the solutions of the VND and the ILS, it represents 540,744 by considering real cost.

MS-ILS Results. The MS-ILS is tested with three runs of the ILS applied with 10 iterations. It has no effect on the instances from the database 1 because the heuristic provides very good solutions. For instances from the database 2, MS-ILS is slightly less effective than VND: 0.175 % is lost as quality of solutions. It is the same when comparing the solutions of MS-ILS and ILS: ILS is better than MS-ILS of 0.234 %. The average calculation time for these instances (databases 2) is very significant: 2590.60 s for the MS-ILS. The method MS-ILS is not interesting for these instances. The MS-ILS provides solutions to 5.349 % to the optimum, for the instances randomly generated. The MS-ILS provides better solutions compared to VND and ILS, for these instances. The improvement of the quality of the solutions is 1.394 % compared to the solutions of the VND and 0.287 % compared to the solutions of the ILS. The average computational time (2245.76 s) is, above those of the VND (292.52 s), and the ILS (964.97 s).

Table 2. Results of VND, ILS (10 iterations) et MS-ILS (3 runs of 10 iterations).

Instances from	VND		ILS		MS-ILS	
	%Gap VND/LB	Time (s)	%Gap ILS/LB	Time (s)	%Gap MS-ILS/LB	Time (s)
Database 1	0.009	2.82	0.009	47.71	0.009	160.00
Database 2	1.184	142.10	1.125	1324.90	1.359	2590.60
Random generator	6.743	292.52	5.636	964.97	5.349	2245.76
Average for all	4.234	202.70	3.567	851.23	3.445	1890.47

For all the evaluated instances, the average gap between the lower bound of MILP resolution and the MS-ILS is 3.445 %. The average computational time of MS-ILS is 1890.47 s.

7   Conclusion

The problem under study is an actual industrial problem, the methods are included in an APS (Advanced Planning System). Some customers reject the CPLEX solution because we can not explain with simple rules how it is built, and require heuristics for which they can understand the logic. The methods VND ILS and MS-ILS have been implemented and tested. The results are compared to those provided by the MILP solver. The VND method significantly improves the quality of the solutions and it is 11 times faster than the MILP solver. The cost of initial solutions (heuristic) decreases by, in average, 17.9 % when the VND is applied (4.234 % of the optimum). The ILS method (with 10 iterations) improves the quality of the solutions (3.567 % of optimum). The average computational time increases, but is below the computation time of the MILP solver. The MS-ILS consumes a lot of time even if it provides, in average, solutions 3.445 % of the optimum. The average computation time of the MS-ILS performed with 3 runs of 10 iterations per ILS) is more than double that of the ILS (with 10 iterations): 1890.47 s for the MS-ILS and 851.23 s for the ILS.

There are some instances for which the solver cannot give a good lower bound within the limited time. Our next step is to develop a Lagrangian relaxation to evaluate these instances with the other methods: greedy heuristic, VND, ILS and MS-ILS. The solutions of Lagrangian relaxation could also be used to build a feasible solution with a repair heuristic.

Acknowledgements. This work was supported by FuturMaster, a french software publisher. The reviewers of the paper are greatly acknowledged for their helpful This comments.

References

1. Cuervo, D.P., Goos, P., Sörensen, K., Arràiz, E.: An iterated local search algorithm for the vehicle routing problem with backhauls. Eur. J. Oper. Res. **237**, 454–464 (2014)
2. Duhamel, C., Lacomme, P., Quilliot, A., Toussaint, H.: A multi-start evolutionary local search for the two-dimensional loading capacitated vehicle routing problem. Comput. Oper. Res. **38**, 617–640 (2011)
3. Feo, T.A., Bard, J.: Flight scheduling and maintenance base planning. Manage. Sci. **35**(12), 1415–1432 (1989)
4. Feo, T.A., Resende, M.G.C.: Greedy randomized adaptive search procedures. J. Global Optim. **6**, 109–133 (1995)
5. Hansen, P., Mladenović, N.: Variable neighborhood search: Principles and applications. Eur. J. Oper. Res. **130**(3), 449–467 (2001)

6. Kande, S., Prins, C., Belgacem, L.: Modèle linéaire mixte et heuristiques pour un réseau de distribution à deux èchelons pour des produits pèrissables. Inf. Syst. Oper. Res. (INFOR) **52**(2), 73–94 (2014)
7. Lourenço, H.R., Martin, O.C., Stützle, T.: Iterated local search. In: Glover, F., Kochenberger, G. (eds.) Handbook of Metaheuristics. International Series in Operations Research and Management Science, pp. 321–353. Kluwer Academic Publishers, Norwell, MA (2002)
8. Lourenço, H.R., Martin, O.C., Stützle, T.: Iterated local search: framework and applications. In: Gendreau, M., Potvin, J.-Y. (eds.) Handbook of Metaheuristics, vol. 146, pp. 363–397. Springer, New York (2010)
9. Michallet, J., Prins, C., Amodeo, L., Yalaoui, F., Vitry, G.: Multi-start iterated local search for the periodic vehicle routing problem with time windows and time spread constraints on services. Comput. Oper. Res. **41**, 196–207 (2014)
10. Nguyen, V.P., Prins, C., Prodhon, C.: A multi-start iterated local search with tabu list and path relinking for the two-echelon location-routing problem. Eng. Appl. Artif. Intell. **25**, 56–71 (2012)
11. Prais, M., Ribeiro, C.C.: Reactive GRASP: an application to a matrix decomposition problem in TDMA assignment. INFORMS J. Comput. **12**(3), 164–176 (2000)
12. Prins C., Wolfler Calvo R.: A fast GRASP with path relinking for the capacitated arc routing problem. In: Gouveia, L., Mouro, C. (eds.) INOC 2005: 3rd International Network Optimization Conference, Lisbonne, Portugal, 20–23 March 2005, pp. 289–295. Université de Lisbonne (2005)
13. Villegas, J.G., Prins, C., Prodhon, C., Medaglia, A.L., Velasco, N.: GRASP/VND and multi-start evolutionary local search for the single truck and trailer routing problem with satellite depots. Eng. Appl. Artif. Intell. **23**, 780–794 (2010)

Critical Activity Analysis in Precedence Diagram Method Scheduling Network

Salman Ali Nisar[(⊠)] and Koji Suzuki

Nagoya Institute of Technology, Nagoya, Japan
ali_nisar02@yahoo.co.jp, Suzuki.koji@nitech.ac.jp

Abstract. Pecedence Diagram Method (PDM) scheduling network introduced three alternative relationships, i.e., finish-to-finish, start-to-start, and start-to-finish, with positive time lag factors between project tasks. PDM with its new relationships has provided more flexible realistic project representation in schedule networks. However, the new relationships of PDM will change the basic concept of critical activities and critical path. Changing the duration of some critical activity in PDM in order to reduce the project duration has anomalous effects on project duration. In this study we proposed classification of critical activity effects on project duration in PDM. In addition, we determined the maximum amount of time by which changing the duration of a critical activity will have anomalous effect on project duration. Therefore, the project managers will clearly distinguish the behavior of each critical activity on critical path, and they can change the project duration by shortening/lengthening activities based on project budget and project deadline.

Keywords: Construction management · Project scheduling · Critical Path Method · Precedence Diagram Method

1 Introduction

Critical Path Method (CPM) is a powerful network diagramming for representation of activities of project. The CPM has been widely used as a construction project management tool to improve scheduling and project administration tasks, and supporting project managers to ensure a project to be completed on time and within the budget [1].

However, the traditional CPM was essentially limited to finish-to-start (FS) relationships between activities, i.e., the successor activity cannot start until the predecessor activity is completed. Therefore, it could not allow overlapping unless activities were further divided [2]. Precedence Diagram Method (PDM), which was developed based on the concept of CPM analysis, introduced three alternative relationships, i.e., start-to-start (SS), finish-to-finish (FF), and start-to-finish; and lag factor between various activities. PDM with three additional relationships between activities has provided a more flexible realistic project representation in schedule networks and more accurately reflects the sequence of construction operations as they occur in real life [1]. There are various computer software packages available such as Primavera P6 and MS Project, which provide the PDM schedule. however, despite the long history and expanding use of PDM, it has serious drawbacks. However, the new relationships of

© Springer International Publishing Switzerland 2015
D. de Werra et al. (Eds.): ICORES 2015, CCIS 577, pp. 232–247, 2015.
DOI: 10.1007/978-3-319-27680-9_14

PDM can change some of the basic concept of critical activities and critical path. According basic definition of critical activity in CPM, the shortening/lengthening of a critical activity on critical path always results in decreased/increased project duration. But, this definition does not always apply on PDM. Changing the duration of some critical activities in PDM will have anomalous effects on project duration [3, 4].

In this paper we classified the critical activity's effect on project duration, and provide important information for critical activity and critical path in PDM. In addition, we determine the maximum amount of time by which changing the duration of a critical activity will have anomalous effect on project duration. Therefore, the project manager will clearly distinguish the behavior of each critical activity on critical path, and he can change the project duration by shortening/lengthening activities based on project budget and project deadline.

2 Scheduling Techniques Overview

2.1 Critical Path Method

Critical Path Method (CPM), which is developed in the 1950 s, is powerful networks diagramming for representation of activities of project and providing project schedule. The CPM generates useful information about the project, such as the shortest project duration, the critical path, and the Total Float (TF) and the Free Float (FF) of each activity. This information is substantially important for the project manager to plan and control the project more actively and efficiently [1, 5]. Critical path and float are the most important concepts among the all information in a project schedule. There are several types of floats, of which the simplest and most important type of float is total float (TF) and free float (FF). TF is the maximum amount of time for which the finish date of an activity can be delayed without affecting the completion of the entire project. TF is calculated as the difference between late start time (LS) and early start time (ES) or between late finish time (LF) and early finish time (EF) of an activity. FF is the amount of time for which the finish date of an activity can be delayed without affecting the start time of any other activities in a project. FF is calculated as the difference between the earliest ES among all the immediate successors of an activity and the EF of that activity. Critical path is the longest ordered sequence of activities through the project schedule, and it determines the earliest time by which a project can be completed. This time is often known as the project duration but more commonly as the critical path. A schedule may have more than one critical path. Each activity in critical path is known as a critical activity. The critical activity has zero TF. When any of them is delayed, it causes a delay in the project completion date.

To perform CPM analysis, the activities that make up the project are first identified. A project network is then used to represent the precedence relationships among activities. Once the project network is drawn the following steps are performed [6].

- A forward pass to determine the ES and LS of the activities,
- A backward pass to determine the EF and LF of the activities,
- Floats calculations, and
- Identification of critical activities and critical path(s).

2.2   Precedence Diagram Method

Precedence Diagram Method (PDM), which is the variation of CPM, has added more flexibility regarding activity relationships while the schedule calculations still utilize as CPM analysis. In PDM, as shown in Fig. 1, an activity can be connected from either its start or its finish, which in addition to the traditional finish-to-start allows the use of three additional relationships between project activities: start-to-start (SS), finish-to-finish (FF), and start-to-finish (SF).

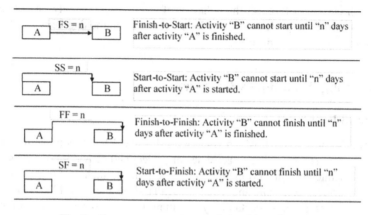

Fig. 1. Four types precedence relationships of PDM.

Another characteristic of PDM diagrams is that periods of time can be assigned between the start and/or finish of one activity and the start and/or finish of a succeeding activity. These periods of time between the activities are referred to as leads and lags. A lead is the amount of time by which an activity precedes the start of its successor, and a lag is the amount of time delay between the completion of one task and the start and/or finish of its successor. Most of commercial software, such as Primavera Project Planner and Microsoft Project allow using the non-traditional relationships with lags (Ahuja 1984). The PDM is also called activity on node (AON) network, and some authors call both methods (CPM and PDM) as CPM.

PDM seems to be more friendly compare to the CPM. For example, suppose that providing concrete floor for a warehouse needs three construction activities. Let activity A is "install formwork, activity B is "reinforcement arrangement", and activity C is "pour concrete". The duration of each activity is estimated as 4, 4 and 2 days, respectively. Based on CPM assumption the technological relationship between these three activities will be finish-to-start, and they have to be executed as series. Therefore, the project duration would be 10 days. Whereas, in PDM as shown in Fig. 2, activity A "reinforce arrangement" and activity B "install formwork" can be executed concurrently so that start-to-start with two days lag would be the prober relationship between them. Thus, the PDM relationship reduces the project duration to 8 days.

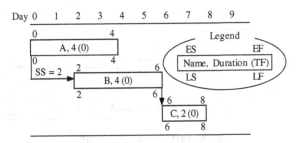

Fig. 2. PDM schedule with SS relationship.

3 Problem Describtion

Despite the long history and expanding use of PDM, it has serious drawbacks. The new relationships of PDM can change some of the basic concept of critical activities and critical path [3, 4]. According basic definition of critical activity in a project schedule, the shortening/lengthening of a critical activity on critical path always results in decreased/increased project duration. But, this definition does not always apply on PDM. Crashing some critical activity in PDM in order to reduce the project duration can have anomalous effects [3, 4].

To better illustrate the anomalous effect of new relationships of PDM on critical activities, consider the simple project schedule that shows on Activity-on-Node network in Fig. 3. Each rectangle in network represents a project activity. The technological relationships between activities are indicated by arrows. The project consists of 7 activities. Activities S and F are assumed to be artificial activities indicating the project commencement and the project completion, respectively. The results of CPM calculation are shown in Fig. 4 as bar-chart fashion along a horizontal time scale. The sequence of activities and precedence arrows denoted by bold line, bold line, represent the critical path. As shown in Fig. 4, we can identify activities 1, 4, 5, 6, and 7 as critical (zero total float). First, let us to define the characteristics of a critical activity in traditional CPM as: (i) any delay in the start time of a critical activity will result in a delay in the project duration, (ii) any change in the length of a critical activity will result the same change in the entire project duration. The first characteristic is true for all critical activities in schedule network in Fig. 4. The second characteristic is still true for critical activities 1, 6, and 7. However, this characteristic cannot be true on critical activities 4 and 5. For example, shortening activity 4 would have reversely effect on project duration, meaning it will increase the project duration. Conversely, lengthening it will decrease the project duration. Also, shortening and lengthening of activity 6 has no effect on project duration. As shown in Fig. 4, these anomalous affects are indicated on critical activities that have SS and/or FF relationships.

In this paper we provide some important information for critical activity effect on project duration. In addition, we determine the maximum amount of time by which changing the duration of a critical activity will have anomalous effect on project duration. Therefore, the project manager will clearly distinguish the behavior of each critical activity on critical path, and he can change the project duration by shortening/lengthening activities based on project budget and project deadline.

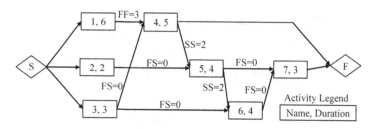

Fig. 3. PDM schedule network.

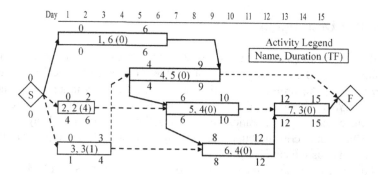

Fig. 4. PDM schedule network on horizontal time scale.

4 Previous Researches

Previous researchers have studied on critical activity and critical path in PDM. Lu and Lam [7] proposed a "transform schemes" in order to detect and transform the new relationships of PDM network i.e., SS, FF, and SF, into equivalent FS relationship by splitting activities. However, some activity may require not to be split during its execution. Therefore, this scheme would not be feasible for such project with non-interruptible activities.

Wiest [4] descried the effects of critical activities with SS and FF relationships. He defined the critical path as alternating sequence of activities and precedence arrows, starting and ending with activities or activity extreme points. Weist [4] classified the critical activities as primary classifications in PDM network as normal, reverse, and neutral.

1. If a critical path passes through an activity from start to finish, then the activity's effect on the critical path or project duration is normal. Its lengthening will increase the critical path, and shortening it will have the opposite effect.
2. If a critical path passes through an activity from finish to start, then the activity's effect on the project duration is anomalous. Its lengthening will shorten the critical path, and its shortening will lengthen the path. Such critical activity is denoted as reverse.

3. If a critical path enters and exits from the starting point of an activity (or ending point), then the duration of activity is independent of the length of that critical path. The activity is called as neutral critical.

Wiest [4] also mentioned that when an activity is on more than one critical path, then the classification would depend on the combination of paths. The combination of normal and reverse is perverse, meaning if one critical path passes through an activity from start to finish and another one enters and exists from its starting point (or ending point), then whether the activity is shortened or lengthened, the project duration will be increased. Such activity is called perverse critical. He also stated that the combination of normal and neutral is normal, the combination of neutral and reverse is reverse, and the combination of all primary classification i.e., neutral, normal, and reverse, will be perverse. However, all these statement are not true. We will show in Sect. 4 that the combination of normal and neutral; and the combination of neutral and reverse will have different effects that Wiest [4] has proposed.

Moder et al. [8] proposed the same classification to the abovementioned. But the only different is that they divided neutral into two classes i.e., start neutral and finish neutral, and also they named the perverse as bicritical.

These classifications provide useful information about the behaviour of critical activities in PDM network, when the project manager needs to change the length of critical activities. However, the previous studies are not completed and could not indicate all the critical activity's characteristics. Therefore, further study is needed in order to provide information in detailed on classification of critical activity effect. In addition, it is needed to determine how long time a certain effect of critical activity would be available during shortening/lengthening of such activity. Because, after shortening/lightening of a critical activity by a certain time unit(s), the activity's effect on project duration may be changed. For example, as the previous classification, activity 5 in Fig. 4 has neutral effect, meaning its lengthening/shortening would not change the project duration. However, when its duration is increased by 2 days, then it will change to normal critical, and then lengthening it will increase the project duration.

5   Network Formulation and Basic Assumption

We consider a single project schedule which is represented by an Activity-On-Node (AON) network in which the activities are denoted by node (circle or rectangle) and the predecessor relationships between predecessor activity i and successor j is shown by an arrow connecting the two nodes. There may exist four types of relationships i.e., SF_{ij}, SS_{ij}, FF_{ij}, and FS_{ij}, with minimum lag time l_{ij} between activities i and j. Each activity is non-preemptive or cannot split during its execution. It is assumed that the resource requirement for each activity is unlimited. The start time of the project is considered as unit time 0. For each activity, the duration and precedence relations are assumed to be deterministic and known in advance.

To avoid having more start activities in the network, an artificial activity with zero duration is used as start activity. If the start of an activity has no predecessor activity, then the start activity is nominated as its direct predecessor with FS relationship.

An artificial activity is used as finish activity. This also helps to have only one finish activity. If the finish of an activity has no successor activity, then the finish activity is used as its direct successor.

6   Classification of Critical Activities

Based on the location of critical activities in the schedule, we classify their effects in two groups i.e., activity on single critical path and activity on multi-critical path. If the critical activity is located on single path, it is called as primary classification by Wiest [4]. But, if it is located on multi-path, then its effect will determine as their combination. The critical path in PDM may define as: the alternating sequence of activities and precedence arrows, starting and finishing with activities or activity extreme points. All the precedence arrows always move forward, and hence an increase in the length (lag time) of a critical precedence arrow will lengthen the project duration. The classifications are described in detailed and depicted in figures as follows. Note that each arrow in the figures represents the critical path direction.

6.1   Critical Activity on Single Path

If the critical activity is located on a single path, then it would classify as primary in 3 classes. The other classification will be provided by combination of these 3 primary classes. In this paper we accept the proposed primary classification by Wiest [4].

1. Normal (N): denotes an activity that lengthening it will lengthen the project duration and shortening it will increase the project duration. Figure 5 show a normal critical activity that critical path *ab* passed through the activity from start to finish.

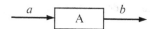

Fig. 5. Normal critical activity.

2. Reverse (R): denotes an activity that lengthening it will shorten the project duration and shortening it will decrease the project duration. Figure 6 depicts the reverse critical activity in which a single critical path of *ab* passed through the activity from finish to start.

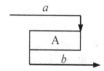

Fig. 6. Reverse critical activity.

3. Neutral (U): denotes an activity that its length is independent of the project duration. There are two types of neutral critical activities. (a) Start-neutral (SU): if a critical path enters and exits from the starting point of an activity, then the activity is called start-neutral critical. (b) Finish-neutral (FU): if a critical path enters and exits from the ending point of an activity, then the activity is called finish-neutral critical. Figure 7 shows the start-neutral and finish-neutral critical activities.

Fig. 7. (A) Start-neutral critical activity, (B) finish-neutral critical activity.

6.2 Critical Activity on Multi-path

If multi-path passes through a critical activity, then the effect of activity on project duration would be depended on the type of combination of paths. Although, Wiest [4] have stated the effect of critical activity when it is located on more than one critical path, all the statement is not true. For example, as Wiest [4] have mentioned that the combination of neutral and reverse will result reverse. However, in following we will show that it would have different result.

1. Perverse (P): denotes an activity which is provided by combination of reverse and normal critical paths. Whether the duration of a perverse critical activity is shortened or lengthened, the project duration will be increased. Figure 8 depicts a perverse critical activity in which the critical path ab (reverse path) enters and exits from finish to start of activity A, while the critical path cd (normal path) enters and exits from start to finish of the activity. Shortening the activity A in Fig. 8 will decrease the path cd, but increase the path ab. And lengthening it will decrease the path ab, but increase the path cd. Therefore, whether activity A is shortened or lengthened, at least one path will increase, and hence the project duration will also increase.

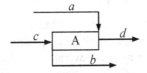

Fig. 8. Perverse critical activity.

2. Decrease-reverse (DR): the combination of neutral (start-neutral or finish-neutral) and reverse will result decrease-reverse effect on project duration. Shortening a decrease-reverse activity will increase the project duration, but lengthening it will have no effect on project duration. For example, shortening activity A in Fig. 9 will

increase the path *ab*, but have no effect on path *cb*. Then, the project duration will increase because the path ab which would be the longest path, will determine the project duration. Lengthening this activity will decrease the path ab, but have no effect on path cb. So, the project duration will not be changed. Because path *cb* would be the longest path and it will determine the project duration. There are two types of DR, i.e. start-decrease-reverse critical activity and finish-decrease-reverse critical activity.

Fig. 9. (A) Start-decrease-reverse critical activity, (B) finish-decrease-reverse critical activity.

3. Increase-normal (IN): the combination of neutral (start-neutral or finish-neutral) and normal will result increase-normal effect on project duration. Lengthening an increase-normal activity will increase the project duration, but shortening it will have no effect on project duration. For example, lengthening activity A in Fig. 10 will increase the path *ac*, but have no effect on path *ab*. Then, the project duration will increase. Shortening the activity will decrease the path ac, but have no effect on path *ab*. Therefore, the project duration will not be changed. Figure 10 shows the start and finish increase-normal critical activities.

Fig. 10. (A) Start-increase-normal critical activity, (B) Finish-increase-normal activity.

7   Determining Float for Critical Activity

7.1   Float for Non-critical Activity

Before identifying float for critical activity in PDM, it is needed to have an observation on traditional definition of float. An activity with positive float is called as non-critical activity. A non-critical activity may have several types of floats i.e., total float (TF), free float (FF), and independent float (IF). Each type of float gives us important information about the characteristic and the flexibility of activity. TF is the maximum amount of time an activity can be delayed from its early start without delaying the entire project. FF is defined as the maximum amount of time an activity can be delayed without delaying the early start of the succeeding activities. FF is the part of TF, hence it is always true that TF ≥ FF. Independent float is the maximum amount of time an activity can be delayed without delaying the early start of the succeeding activities and without being affected by the allowable delay of the preceding activities [9].

So far, all floats are associated with non-critical activity. In this study we introduce some new types of float associated with critical activity. These float would concretely define the characteristic of a critical activity in PDM network.

7.2   Float for Critical Activity

An activity with zero TF is denoted as critical activity. Obviously if TF of an activity is zero, then its FF and IF will be also zero. Because FF and IF are the parts of TF [9]. However, a critical activity in PDM may have several type of floats other than TF, FF, and IF.

1. Reverse Float (RF): it is the maximum amount of time associated with a reverse critical activity that when the length of activity is extended, then critical path will be decreased. After consuming the entire reverse float, the activity effect will be changed to decrease-critical. For example, lengthening activity 4 (which is a reverse critical activity) in Fig. 4 by 1 day will shorten the project duration from 15 to 14 days. Afterward lengthening it would have no anymore effect on project duration, and it would be changed to decrease reverse activity. Therefore, this activity has 1 day reverse float. The reverse float is calculated from Eq. 1.

$$RF_R = min \begin{cases} ES_R - EF_i - l_{iR}, \forall FS_{iR} \\ ES_R - ES_i - l_{iR}, \forall SS_{iR} \\ LS_j - EF_R - l_{Rj}, \forall FS_{Rj} \\ LF_j - EF_R - l_{Rj}, \forall FF_{Rj} \end{cases} \qquad (1)$$

where, R is the reverse critical activity, RF_R is reverse float of activity R, i is the predecessor activity, and j is the successor activity.

2. Neutral Float (UF): it is the maximum amount of time associated with neutral critical activity by which the length of activity can be extended without affecting the duration of critical path. After use of all neutral float, the activity effect will be changed to normal critical. For example, activity 5 in Fig. 4 is neutral critical so that lengthening it by 2 days will have no effect on project duration. But, lengthening it by more than 2 days will increase the project duration. Thus, activity 6 has 2 days of UF. The UF is calculated from Eqs. 2 and 3
For finish-neutral critical activity:

$$UF_U = min \begin{cases} LS_j - EF_U - l_{Uj}, \forall FS_{Uj} \\ LF_j - EF_U - l_{Uj}, \forall FF_{Uj} \end{cases} \qquad (2)$$

For finish-neutral critical activity:

$$UF_U = min \begin{cases} ES_U - EF_i - l_{iU}, \forall FS_{iU} \\ ES_U - ES_i - l_{iU}, \forall SS_{iU} \end{cases} \qquad (3)$$

where, U is the neutral critical activity and UF_U is neutral float of activity U.

3. Decrease-reverse Float (DF): it is the maximum amount of time associated with decrease-reverse critical activity by which lengthening the activity would not effect on project duration. After consumption of DF, the activity effect will be changed to perverse critical. For example, let us consider on the simple schedule network in Fig. 10 in which the activity 3 is a decrease-reverse critical. When the duration of activity 3 is lengthened by 2 days, it would have no effect on project duration. But, if it is lengthened by 3 days (or more than 2 days), then the project duration will be increased. Thus, activity 3 in Fig. 11 has 2 days DF. The DF is calculated from Eqs. 4 and 5.

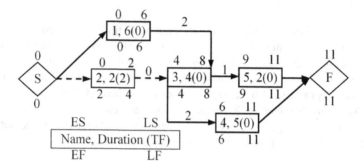

Fig. 11. Decrease-reverse critical activity in example schedule.

For start-decrease-reverse critical activity:

$$DF_D = min \left\{ \begin{array}{l} LS_j - EF_D - l_{Dj}, \forall FS_{Dj} \\ LF_j - EF_D - l_{Dj}, \forall FF_{Dj} \end{array} \right\} \tag{4}$$

For finish-decrease-reverse critical activity:

$$DF_D = min \left\{ \begin{array}{l} ES_D - EF_i - l_{iD}, \forall FS_{iD} \\ ES_D - ES_i - l_{iD}, \forall SS_{iD} \end{array} \right\} \tag{5}$$

where, D is the decrease-reverse critical activity and DF_D is decrease-reverse float of activity D.

8 Case Study

To better illustrate the critical activity classification and identifying the proposed floats for each class, we use a slightly more complicate example which is taken from text of Wiest and Levy [3]. It is assumed that there is the construction of a large condominium project, in which a number of essentially identical housing units are built sequentially. The example follows three units only and is concerned with just the laying of cement slabs. Performing laying of cement slab for each unit is broken down into 5 activities as follows.

(1) Clear lot and grade 8 h
(2) Place concrete forms 12 h
(3) Lay sewer lines 16 h
(4) Install reinforcement steel 9 h
(5) Pour concrete and smooth 4 h

Total number of activity that must be performed for 3 units is 15 activities. These activities are numbered sequentially, unit by unit as: activities 1 through 5 refer to unit 1, activities 6 through 10 refer to unit 2, and activity 11 through 15 refer to unit 3. For example, activity 2 is "Place concrete forms, unit 2" and activity 14 is "Install reinforcement steel, unit 3." The example data with proper precedence relationships are shown in Table 1. Figure 11 represents the schedule network which is drawn on Activity-On-Node fashion, with precedence arrows connecting the activities at the appropriate ends (start or finish). The time lag factor (l_{ij}) is inside the box attached to arrows. Using the forward and backward calculations of PDM algorithm, as shown in Table 2, we calculated the ES, EF, LS, LF, TF, and the project duration. The project duration is 56 working hours. We can observe the critical activities (activity with zero TF) in Table 2 as activities 1, 2, 3, 7, 8, 12, 13, 14, and 15. These activities are denoted by bold numbers in Table 2. As shown in Fig. 11, the sequences of critical activities and critical precedence arrows that are highlighted by bold line indicated the critical paths. To better illustrate, we construct the critical paths separately in Fig. 13. As shown in Fig. 13, we identified 3 following critical paths: (1) the sequences activities of 1, 3, 8, 13, 12, 14, and 15 connecting by precedence relationships of $FS_{1,3}$, $SS_{3,8}$, $SS_{8,13}$, $FF_{13,12}$, $SS_{12,14}$, $FS_{14,15}$; (2) the sequences activities of 1, 3, 2, 7, 12, 14, and 15

Table 1. Activity information for laying cement slabs.

Activity	Description	Duration	Precedence Relationship
1	Clear & Grade 1	8	
2	Con. Forms 1	12	1 FS, 3 FF=2
3	Sewe Lines 1	16	1 FS
4	Reinforcement 1	9	2 FF=1 & SS=7, 3 SS=7
5	Pour Concrete 1	4	4 FS=2
6	Clear & Grade 2	8	1 FS
7	Con. Forms 2	12	2 SS=10, 6 FS, 8 FF & SS=7
8	Sewe Lines 2	16	3 SS=10, 6 FS
9	Reinforcement 2	9	4 FS, 7 SS=7 & FF=1, 8 SS=7 & FF
10	Pour Concrete 2	4	5 SS=2, 9 FS=2
11	Clear & Grade 3	8	6 FS
12	Con. Forms 3	12	7 SS=10, 11 FS, 13 FF=2
13	Sewe Lines 3	16	8 SS=10, 11 FS
14	Reinforcement 3	9	9 FS, 12 SS=7 & FF=1, 13 SS=7 & FF
15	Pour Concrete 3	4	10 SS=2, 14 FS=2

Table 2. Schedule information by traditional PDM algorithm.

Activity	ES	EF	LS	LF	TF
1	0	8	0	8	0
2	14	26	14	26	
3	8	24	8	24	0
4	21	30	23	32	2
5	32	36	44	48	12
6	8	16	10	18	2
7	24	36	24	36	0
8	18	34	18	34	0
9	31	40	32	41	1
10	42	46	50	54	8
11	16	24	20	28	4
12	34	46	34	46	0
13	28	44	28	44	0
14	41	50	41	50	0
15	52	56	52	56	0

connecting by precedence relationships of $FS_{1,3}$, $FF_{3,2}$, $SS_{2,7}$, $SS_{7,12}$, $SS_{12,14}$, $FS_{14,15}$; and (3) the sequences activities of 1, 3, 8, 7, 12, 14, and 15 connecting by precedence relationships of $FS_{1,3}$, $SS_{3,8}$, $FF_{8,7}$, $SS_{7,12}$, $SS_{12,14}$, $FS_{14,15}$. The duration of critical path in PDM will be calculated from following equation.

$$d_{Pi} = \sum d_{Ni} + \sum l_{ij} - \sum d_{Ri} \tag{6}$$

where, d_{Pi} is the duration of path, d_{Ni} the duration of normal activity, l_{ij} is the duration lag factor, and d_{Ri} is the duration of reverse activity.

For example, the duration of critical path 1 is calculated as:

$$\sum d_{Ni} = (d_1 + d_{13} + d_{14} + d_{15}) = 37$$

$$\sum l_{ij} = (l_{3,8} + l_{8,13} + l_{13,12} + l_{12,14} + l_{14,15}) = 31$$

$$\sum d_{Ri} = (d_{12}) = 12$$

$$d_{P1} = 37 + 31 - 12 = 56$$

As shown in Table 2, the traditional PDM algorithm provided information for all activities, e.g., critical activities, critical paths, ES, EF, LS, LF, and TF. Since, changing the duration of critical activities in PDM schedule have unusual effects on project duration, this information cannot indicate all the characteristics of critical activities.

The proposed approach in Table 3 provides the completed information for activities including the critical activity classification and critical activity float. As shown in

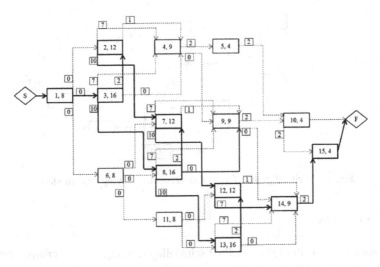

Fig. 12. PDM schedule network of laying cement slabs.

column 8 of Table 3, activities 1, 13, and 15 are normal (N); activity 2 is reverse (R); activity 14 is start-neutral (SU); activity 7 is decrease-reverse (DR); and activities 3, 8, and 12 are increase-normal (IN). An observation should be noted about activity 12 in Fig. 12. The activity 12 has reverse effect on critical path 1, and it has neutral effect on critical paths 2 and 3. However, it would have decrease-reverse effect on project duration. Using the proposed Eqs. 1–5 described earlier, we provided floats for critical activities i.e., reverse float (RF), neutral float (NF), and decrease-reverse float (DF), in columns 7 to 9 of Table 3.

Table 3. Schedule information by proposed approach in PDM.

Activity	ES	EF	LS	LF	TF	RF	UF	DF	Class
1	0	8	0	8	0	–	–	–	N
2	14	26	14	26	0	3	–	–	R
3	8	24	8	24	0	–	–	–	IN
4	21	30	23	32	2	–	–	–	
5	32	36	44	48	12	–	–	–	
6	8	16	10	18	2	–	–	–	
7	24	36	24	36	0	–	–	3	DR
8	18	34	18	34	0	–	–	–	IN
9	31	40	32	41	1	–	–	–	
10	42	46	50	54	8	–	–	–	
11	16	24	20	28	4	–	–	–	
12	34	46	34	46	0	–	–	3	DR
13	28	44	28	44	0	–	–	–	N
14	41	50	41	50	0	–	–	–	N
15	52	56	52	56	0	–	–	–	N

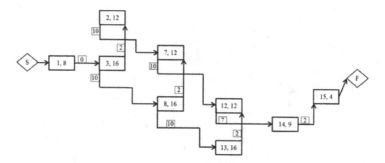

Fig. 13. Critical paths of schedule network of laying cement slabs.

9 Conclusions

Traditional Critical Path Method (CPM) scheduling technique is essentially limited to finish-to-start (FS) relationships between activities, i.e., the successor activity cannot start until the predecessor activity is completed. Therefore, it could not allow over-lapping unless activities were further divided. Precedence Diagram Method (PDM), which was developed based on the concept of CPM analysis, introduced three alter-native relationships, i.e., start-to-start (SS), finish-to-finish (FF), and start-to-finish; and lag factor between various activities. The PDM provides a more flexible realistic project representation in schedule networks and more accurately reflects the sequence of construction operations as they occur in real life. There are various computer software packages available such as Primavera P6 and MS Project, which provide the PDM schedule network. However, the new relationships of PDM can change some of the basic concept of critical activities and critical path. According basic definition of critical activity in CPM, the shortening/lengthening of a critical activity on critical path always results in decreased/increased project duration. But, this definition does not always apply on PDM. Because changing the duration of some critical activity in PDM will have anomalous effects on project duration. For example, when the duration of critical activity in PDM schedule is shortened, then the project duration may increase rather than decrease. This situation is a significant problem for project managers because in order to complete the project before the project deadline, sometime they have to reduce the project duration by crashing the critical activities on critical path. Since changing the duration of critical activity in PDM has anomalous effects on project duration, the project managers cannot decide that the shortening of which critical activity will reduce the project duration. Therefore, further research is needed to provide more information about the critical activity in PDM network.

Previous researchers have studied on critical activity and critical path, and they have proposed classification of critical activity. However, these classifications were not completed and could not indicate all the critical activity's characteristics. Therefore, in this paper we did further research on classification of critical activity effects and introduced new floats for each class of critical activities in PDM schedule. We clas-sified the critical activity's effect in 6 classes, i.e., normal, reverse, neutral, perverse, decrease-reverse, and increase-normal. This classification is completed and indicates all the effects of critical activities on project duration when their duration is changed. Furthermore, we introduced 3 new types of floats that are revers float, neutral float, and decrease-reverse float for reverse, neutral, and decrease-reverse, respectively. These floats determine that how much an effect is true on a certain critical activity. This research will help the project managers to clearly distinguish the behavior of each critical activity on critical path, and to can change the project duration by shortening/lengthening activities based on the project budget and the project deadline.

References

1. Ahuja, H.N., Dozzi, S.P., Abourizk, S.M.: Project Management Techniques in Planning and Controlling Construction Projects, 2nd edn. Wiley, New York (1994)
2. Fondahl, J.W.: A non-computer approach to the critical path method for the construction industry. J. Constr. Eng. Manage. 132(7), 697–711 (1961)
3. Wiest, D.J., Levy, K.F.: A management guide to PERT/CPM with GERT/DCPM and other networks, 2nd edn. Prentice-Hall Inc, New Jersey (1977)
4. Wiest, J.D.: Precedence diagramming method: some unusual characteristics and their implications for project managers. J. Oper. Manage. 1(3), 121–130 (1981)
5. Bowers, J.A.: Criticality in resource-constrained networks. J. Oper. Res. Soc. 46(1), 80–91 (1995)
6. Hegazy, T.: Computer-Based Construction Project Management. Prentice Hall, Upper Saddle River (2002)
7. Lu, M., Lam, H.C.: Transform schems applied on non-finish-to-start logical relationships in project network diagrams. J. Constr. Eng. Manage. 135(9), 863–873 (2009)
8. Moder, J.J., Phillips, C.R., David, E.W.: Project Management with CPM, PERT and Precedence Diagramming, 3rd edn. Van Nostrand-Reinhold, New York (1983)
9. Mubarak, S.: Construction Project Scheduling and Control, 2nd edn. Wiley, Hoboken (2010)

Author Index

Printed in the United States
By Bookmasters